GLUCONEOGENESIS

Edited by **Weizhen Zhang**

Gluconeogenesis
http://dx.doi.org/10.5772/63700
Edited by Weizhen Zhang

Contributors

Leszek Szablewski, Kirill Vitoldovich Pozhar, Nikolay Bazaev, Richa Karmakar, Zhuo Mao, Weizhen Zhang

Published by InTech

Janeza Trdine 9, 51000 Rijeka, Croatia

Publishing Process Manager Martina Usljebrka
Technical Editor SPi Global
Cover InTech Design team

First published July, 2017
Printed in Croatia
Legal deposit, Croatia: National and University Library in Zagreb

Additional hard copies can be obtained from orders@intechopen.com

Gluconeogenesis, Edited by Weizhen Zhang
p. cm.
Print ISBN 978-953-51-3323-0
Online ISBN 978-953-51-3324-7

For EU product safety concerns:
IN TECH d.o.o., Prolaz Marije Krucifikse Kozulić 3, 51000 Rijeka, Croatia,
info@intechopen.com or visit our website at intechopen.com.

Meet the editor

Weizhen Zhang is a professor of Physiology. Dr. Zhang received his bachelor's degree in Medicine from Sun Yat-sen University in 1987 and PhD degree from Peking University in 1994. Dr. Zhang has published over 100 peer-reviewed papers in journals such as PNAS, Diabetes, Diabetologia, Endocrinology, and Molecular Endocrinology. Other recognitions include 10 book chapters, 20 professorships, 30 invited presentations, ad hoc reviewer for over 30 peer-reviewed journals, editorial board members of 6 peer-reviewed journals, and standing members of grant review committees such as NSFC and ADA. Dr. Zhang's research focuses on defining the mechanisms that underlie fuel sensing in the gastrointestinal tract and how the fuel sensor in GI tract coordinates the whole-body energy status to control energy metabolism.

Contents

Preface

Glucose homeostasis is intricately regulated in human body. Dysregulation of glucose homeostasis results in a number of diseases or multiorgan complications. Gluconeogenesis, the metabolic process by which glucose or glycogen is synthesized from noncarbohydrate substrates, is especially important for maintaining the plasma glucose level within normal range during the fasting or nutrition deprivation state. In the past several decades, the knowledge of gluconeogenesis has been rapidly growing. This book aims to provide an overview on the topic of gluconeogenesis, its regulation, and impact on the glucose homeostasis. The authors discuss gluconeogenesis from different aspects and hope to enhance a clear understanding of this critical and complex biochemical process.

This book mainly contains four chapters focusing on the topic "gluconeogenesis." Chapter 1 begins with an overview of carbohydrate and its transport mechanisms using different transporters. General knowledge of glucose metabolism is then reviewed, including glycolysis, oxidative decarboxylation, Krebs cycle, electron transport chain, glycogenesis, glycogenolysis, and gluconeogenesis. The roles of peripheral tissues and hypothalamic-pituitary axis in glucose homeostasis are also addressed. Chapter 2 introduces the carbohydrate metabolism in a simple eukaryotic microorganism called *Dictyostelium discoideum*. It presents the metabolic details of the gluconeogenesis in eukaryotic cells and addresses the developmental stages of *Dictyostelium discoideum*. In the end, how gluconeogenesis process affects the size of the aggregates and differentiation stages is also discussed. Chapter 3 first outlines the definition of "gut-brain-liver axis" and discusses the gut-DVC-liver axis and the gut-hypothalamus-liver axis, respectively. It then discusses the regulation of glucose homeostasis by distinct gut hormones, such as GLP-1 and ghrelin. The final section addresses the importance of hypothalamic hormonal signaling and nutrient sensing in glucose metabolism. Chapter 4 focuses on the application of bioinformatics and computational biology in artificial pancreas. The author addresses modeling of blood glucose dynamics, development of blood glucose prediction algorithm, and its approbation on clinical data.

I am very grateful for the contributions from these outstanding scientists and hope that this book will stimulate all scholars to dedicate their research in revealing the relevant mechanism and its application of gluconeogenesis. I would also like to thank Ms. Andrea Korić and Ms. Ušljebrka for their great efforts in the book planning and editing during the process of book publication.

Dr. Weizhen Zhang
Professor of Physiology
Department of Surgery
University of Michigan Medical Center
USA

Chapter 1

Introductory Chapter: Gluconeogenesis

Weizhen Zhang

Additional information is available at the end of the chapter

http://dx.doi.org/10.5772/intechopen.69801

Glucose, a prominent metabolic fuel in mammals, is the precursor for synthesis of all the other carbohydrates, such as glycogen, ribose, glycolipids, and glycoproteins. Blood glucose levels need to be maintained within a relatively narrow range, which is defined as the glucose homeostasis. This stable equilibrium of blood glucose levels is achieved by a balance between the dietary glucose uptake, endogenous glucose production, glucose partition, and glucose utilization. Derangement of blood glucose could cause severe outcomes. Hypoglycemia may lead to brain dysfunction, and if more severe, even death. Acute hyperglycemia sometimes reflects stressed clinical conditions such as injury or infection with poor outcomes. Chronic elevation of blood glucose causes diabetes, which results from impairment of organism glucose homeostasis.

The contribution of endogenous glucose production includes gluconeogenesis and glycogenolysis. Gluconeogenesis is the metabolic process in which glucose is produced from certain noncarbohydrate carbon substrates. These substrates include the breakdowns of proteins and lipids such as amino acids, lactate, and glycerol. Liver and kidney are two important tissues of gluconeogenesis, which are responsible for maintenance of normal blood glucose level in the fasting state or starvation. Although small intestine also expresses the key gluconeogenic enzymes, it is still not clear whether this organ produces glucose in the fasting state. Insulin and glucagon are two counteracting hormones critical for the regulation of gluconeogenesis. These hormones can act acutely to alter the enzyme activity or chronically through gene expression. In addition to the peripheral targets, these hormones have been recently identified to also act through the hypothalamic neurons. Hypothalamic centers involved in the control of endogenous glucose production can constantly sense fuel availability by integrating inputs from circulating nutrients and gluconeogenic hormones. In response to these peripheral signals, the hypothalamus sends out efferent impulses to alter hepatic gluconeogenesis, thus keeping blood glucose levels in the normal range. Disruption of this intricate neural control may contribute to defects of glucose homeostasis and insulin resistance in type 2 diabetes and obesity [1].

This book aims to provide an overview on the gluconeogenesis, its regulation, and impact on the glucose homeostasis. The first chapter addresses the fundamental process by which glucose homeostasis is maintained. In this chapter, mechanisms by which glucose is transported into cells are first reviewed. The first rate-limiting step of glucose metabolism is the transport of monosaccharides into cells. This is achieved by three classes of glucose transporters: the facilitative glucose transporters, the sodium-glucose co-transporters, and sweet family, also known as PQ-loop, Saliava or MtN3 family. The second section provides a comprehensive summary on the critical biochemical processes of glucose metabolism within cells with focus on gluconeogenesis. The roles of distinct tissues such as pancreas, liver, kidney, and hypothalamic-pituitary axis in the glucose homeostasis are fully discussed in the last section. In addition to the classical hormones, insulin and glucagon, pancreatic islet cells also secrete somatostatin, amylin, and pancreatic polypeptide (PP). All these hormones are critically involved in the regulation of glucose metabolism in three major metabolic tissues including liver, adipose tissue, and skeletal muscle. Glucose transport and metabolism, in particular gluconeogenesis, in liver and kidney are comprehensively addressed. The hypothalamic-pituitary-adrenal (HPA) axis in the control of glucose homeostasis and its relation to the regulation of reproduction are reviewed at the end.

The second chapter provides a comprehensive overview on the biochemical process of gluconeogenesis in eukaryotic cells using the *Dictyostelium discoideum* as a model. This chapter opens with the introduction on the general biochemical process of gluconeogenesis in the eukaryotic cells. This is followed by an overview on how gluconeogenesis affects the developmental stages of *D. discoideum* and the differentiation process of *Myxamoebae.* Grown as single cells, *D. discoideum* develops as multicellular organisms. The metabolic pathway and its signaling regulation are similar to that presented in plants and animals. *D. discoideum* is thus the best biological system to study the molecular pathways like glycolysis or gluconeogenesis relative to the mammalian system.

The third chapter provides a comprehensive review on the gut-brain-liver axis and its role in the control of glucose homeostasis. The current knowledge on the gut-brain interaction and its precise control on hepatic gluconeogenesis is provided, with focus on the gastrointestinal hormones and hypothalamic neuronal signaling. In this chapter, the gut-dorsal vagal complex-liver axis and the gut-hypothalamus-liver axis are first introduced. The molecular mechanism underlying the central neuronal regulation of hepatic gluconeogenesis is addressed. As the regulatory center for glucose homeostasis, hypothalamic neurons integrate peripheral hormone signals and nutrient levels to coordinate the hepatic glucose production. As proposed by Rojas and Schwartz [2] in the two-compartment model, the hypothalamus may thus be considered as another key organ critical for glucose homeostasis in addition to the pancreatic islets.

The fourth chapter addresses the new mathematical model for monitoring and prediction of blood glucose. It provides a comprehensive overview on the mathematical modeling of blood glucose dynamics, development of blood glucose prediction algorithm, and its approbation on clinical data. This predictive algorithm is based on the Sigma-model. It estimates the difference between theoretical and experimental tracks with possibility of further correction. Blood glucose prediction algorithm allows for detection of incorrect measurements and correction spikes on blood glucose track, allowing patients to avoid incorrect insulin infusion or glucose intake. In addition, the algorithm can alert about closed-loop blood glucose control system failures. It allows patients to detect damage or cross-clamping of insulin pump infusion set and actualize meal data.

Author details

Weizhen Zhang

Address all correspondence to: weizhenzhang@bjmu.edu.cn

The University of Michigan Medical Center, Michigan, USA

References

[1] Demuro G, Obici S. Central nervous system and control of endogenous glucose production. Current Diabetes Reports. 2006;**6**(3):188-193

[2] Rojas JM, Schwartz MW. Control of hepatic glucose metabolism by islet and brain. Diabetes Obes Metab. 2014;16 (Suppl 1):33-40.

Chapter 2

Glucose Homeostasis

Leszek Szablewski

Additional information is available at the end of the chapter

http://dx.doi.org/10.5772/67222

Abstract

Glucose is the main and preferred source of energy for mammalian cells. Mammalian cells need glucose constantly. Long-lasting disturbances in blood glucose concentrations can cause diseases and death. Therefore, blood glucose concentrations must be within narrow limits. The process of maintaining blood glucose at a steady-state level is called glucose homeostasis.

Keywords: glucose homeostasis, glucose metabolism, pancreas, liver, kidney, hypothalamic-pituitary axis

1. Introduction

Carbohydrates play several roles in the metabolic processes and as structural elements of living organisms. An essential substrate for all mammalian cells is monosaccharide, glucose. In human glucose is obtained directly from the diet or by synthesis in the liver and kidney. Monosaccharides are transported across the intestinal wall to the portal vein and then to liver cells and other tissues. Monosaccharides play a role as a precursor of fatty acids, amino acids, and glycogen. They are also oxidized by the various catabolic pathways of cells, play an important role in the synthesis of pentose sugars for DNA and RNA, as well as generate NADPH + H^+ for the synthesis of lipids.

2. Transport of monosaccharides into cells

In the first step of carbohydrate metabolism, monosaccharides are transported across the plasma membranes. Due to hydrophilic nature of glucose and other monosaccharides, the lipid bilayer of plasma membrane is impermeable for these substances. Therefore, monosaccharide transport across the plasma membrane is mediated *via* membrane transport proteins called glucose transporters.

In human, there are three classes of glucose transporters: the facilitative glucose transporters, the sodium-glucose cotransporters, and SWEETs. However, these transporters are named "glucose transporters"; they transport not only glucose but also other substances such as fructose, fucose, xylose, vitamins, ions, etc. GLUT1 was also suggested to be receptor for human T-lymphotropic virus (HTLV) and plays an essential role in CD4 T-cell activation

2.1. The GLUT family

GLUT proteins are encoded by the *SLC2* genes. These proteins are members of the major facilitator superfamily (MFS) of membrane transporters. These transporters are uniporters. They facilitate the diffusion of substrates across cellular membranes along a concentration gradient [1, 2]. The GLUT family comprises 14 isoforms GLUT1–GLUT12, GLUT14, and HMIT (GLUT13). HMIT is the proton-driven myoinositol transporter.

The human GLUT proteins are comprised of about 500 amino acid residues. They are predicted to possess 12 transmembrane-spanning α-helices and a single N-linked oligosaccharide [4]. The cytoplasmic domain contains a short N-terminal segment, a large intracellular loop between transmembrane domains 6 and 7, and a large C-terminal segment. The sequences among members of family are 14–63% identical and 30–79% conservative [5]. Sequence alignments of all members reveal several highly conserved structures [5]. The fact that the transmembrane domain primary structure is largely conserved suggests that the glucose channel is basically identical in structure among the members of this family [6]. One or more GLUT proteins are expressed in every cell type of the human body. It is highly likely that the major substrates for several GLUT proteins have not yet been identified [3].

2.2. The Na^+/glucose cotransporters

Sodium-glucose transporters, also known as Na^+/glucose cotransporters or symporters, are encoded by *SLC5* genes. These transporters are members of a larger gene family of sodium/substrate symporter family (SSSF) that contain a common SSF motif in the fifth transmembrane region [7]. There are 12 human genes in the *SLC5* family that are expressed in different tissues. These cotransporters transport substrates *via* a secondary active transport mechanism. SGLTs do not directly utilize ATP to transport glucose against its concentration gradient; rather, they must rely on the sodium concentration gradient generated by the sodium-potassium ATPase as a source of chemical potential [8]. The Na^+-electrochemical gradient provided by the Na^+-K^+ ATPase pump is utilized to transport substrate into cells against its concentration gradient. Except for SGLT3 which is glucose sensor, all are sodium cotransporters [9]. The function of ten are tightly coupled plasma membrane Na^+/substrate cotransporters for different solutes, one is a Na^+/Cl^-/Choline cotransporter and one is a glucose-activated ion channel [7].

All members of the *SLC5* family code for 60 to 80 kDa proteins contain 580–718 amino acids [10]. The genes *SLC5* contain 14–15 exons; however, *SLC5A7* gene contains 8 exons, and

http://dx.doi.org/10.5772/67222

SLC5A3 gene contains 1 exon [7]. SGLTs contain 14 transmembrane α-helices (TMH) with both NH_2 terminus and the COOH terminus facing the extracellular (luminal) side of the cell [11, 12]. The human sodium cotransporters share an amino acid identity of 21–70%. The transporter contains a single glycosylation site [11]. Of note, glycosylation is not required for functioning of the protein. Phosphorylation sites are suggested between transmembrane helices 5 and 6 [13] and between transmembrane helices 8 and 9 [14].

2.3. The SWEET proteins

Sugar efflux transporters are essential for the maintenance of human blood glucose levels. In mammals, glucose efflux from the liver is crucial for the maintenance of blood glucose levels. Chen et al. [15] identified a new class of sugar transporters, named SWEETs. The SWEETs are ubiquitously expressed in plant. SWEET belongs to a novel transporter family with 17 members in *Arabidopsis* and 21 in rice [15]. Homologs of SWEETs have also been identified in humans (SWEET1) [15]. Although human and animal genomes typically contain only a single SWEET gene, a major exception is *Caenorhabditis elegans*, which contains seven SWEET paralogs [16]. SWEET is a glucose uniporter.

The human SWEET1 is expressed in the oviduct, epididymis, intestine, and β-cell lines [15]. It is a candidate for the vesicular efflux from enterocytes, hepatocytes, and β cells [15]. The SWEET class of transporters is predicted to have seven transmembrane helices. Eukaryotic SWEETs have a predicted topology comprising a repeat of three membrane-spanning domains that are connected by an inversion linker helix with extracellular N-terminus and intracellular C-terminus [15–17]. Cytosolic C-terminus of the SWEETs is very long and may serve as a docking platform for protein interactions [18]. C-termini show much less conservation and are characterized by extensive length variability [19].

3. General information on glucose metabolism

3.1. Glucose as a source of cellular energy

When energy is needed, glucose is rapidly metabolized to produce adenosine triphosphate (ATP), a high-energy product. Glucose is oxidized through a long series of reactions that extract the great amount of possible energy from it.

3.1.1. Glycolysis

The first which begins the complete oxidation of glucose is called glycolysis (or Embden-Meyerhof-Parnas pathway). It is an anaerobic process. During glycolysis, each glucose molecule is split and converted to two three-carbon units. The next result of glycolysis is the production of two pyruvate molecules, two ATP, and two NADH + H^+. During glycolysis, the cell receives about 5% of the total energy available. In the presence of oxygen, aerobic organisms oxidize pyruvate to CO_2 and H_2O. In the absence of oxygen, pyruvate can be converted to several types of reduced molecules, such as ethanol (e.g., yeast) or lactate (e.g., muscle cells, red blood cells). This anaerobic process is referred to as fermentation.

3.1.2. *Oxidative decarboxylation*

During aerobic metabolism of glucose, pyruvate is transported inside mitochondria, where is oxidized. Oxidation of pyruvate to acetyl coenzyme A (Acetyl-CoA) produces two molecules of CO_2 and NADH + H^+.

3.1.3. *Krebs cycle*

Further series of reactions is collectively called "Krebs cycle," also known as the "citric acid cycle" or the "tricarboxylic acid cycle (TAC)." Through a series of reactions in mitochondria, acetyl-CoA is oxidized to CO_2. Fuel for the Krebs cycle comes also from lipids (fats) and proteins (amino acids), which produce the molecule acetyl-CoA. If carbohydrates are the fuel for Krebs cycle, this cycle occurs twice since each glucose produces two pyruvates and then in the process of oxidative decarboxylation two molecules of acetyl-CoA.

3.1.4. *Electron transport chain*

The oxidative phosphorylation, which occurs in mitochondria, is a series of reactions that utilize the energy from NADH + H^+ and $FADH_2$. The electrons are successively passed down the chain of cytochromes, each time releasing some of their energy, which is then used to pump protons actively across the membrane into the matrix down this chemiosmotic gradient but can only do so through the ATP synthase. The result of the electron transport chain is three molecules of ATP, if a donor of protons and electrons is NADH + H^+ and one molecule of H_2O. In the case of $FADH_2$, the result of this process is two molecules of ATP and one molecule of H_2O.

3.2. Glycogenesis

Glycogenesis is the process of glycogen synthesis from glucose. Glycogen is the storage form of glucose. Glycogenesis occurs after a meal, when blood glucose levels are high. All cells contain glycogen, but most is stored in liver cells (about 90 g in a 70-kg man) and muscle cells (about 350 g in a 70-kg man). In this process, glucose molecules are added to chains of glycogen for storage in mentioned organs.

Glucose-6-phosphate is converted to glucose-1-phosphate by phosphomutase. Sugar-nucleotide synthesis is a reaction preceding sugar polymerization processes. Uridine diphosphate glucose (UDP-glucose) is more reactive than glucose. By itself, this is a readily reversible reaction; however, the subsequent hydrolysis of pyrophosphate to two inorganic phosphates (PPi) will readily occur, and this will drive the reaction over the product side.

For the synthesis of glycogen, the starting point is the protein glycogenin. If the chain contains more than ten molecules of glucose residues, it acts as a primer for proglycogen synthase which elongates primer. The elongation is due to the addition of new glucose molecules to the existing chain.

3.3. Glycogenolysis

When the blood sugar levels fall, glycogen stored in the muscle and liver may be broken down. This process is called glycogenolysis. The liver can consume glucose-6-phosphate in glycolysis and can also remove the phosphate group using the enzyme glucose-6-phosphatase and release the free glucose into the bloodstream. Since muscle cells lack glucose-6-phosphatase, they cannot convert glucose-6-phosphate into glucose and therefore use the glucose-6-phosphate to generate energy for muscle contraction.

3.4. Gluconeogenesis

Gluconeogenesis generates glucose from noncarbohydrate precursors such as lactate, glycerol, pyruvate, and glucogenic amino acids. It occurs primarily in the liver. Under certain conditions, such as metabolic acidosis or starvation, the kidney can make small amounts of new glucose. When liver glycogen is depleted, the gluconeogenesis pathway provides the body with adequate glucose. The major substrates for gluconeogenesis are lactate (formed in muscle and red blood cells), amino acids (derived from the muscle), and glycerol (produced from the degradation of triacylglycerols). During anaerobic glycolysis, pyruvate is reduced to lactate. Lactate is released to the bloodstream and transported into the liver. In the liver lactate is converted to glucose, and then glucose is returned to the blood for use by the muscle as an energy source. This cycle is termed the Cori cycle. The gluconeogenesis of the cycle is a net consumer energy, costing the body four molecules of ATP more than are produced during glycolysis. The reaction sequence in gluconeogenesis is largely the reverse of glycolysis.

Of all the amino acids that can be converted to glycolytic intermediates, alanine is perhaps the most important. When the muscle produces large quantities of pyruvate, for example, during exercise, some of these molecules are converted to alanine. Alanine is transported to the liver, reconverted to pyruvate and then to glucose. This cycle is termed the glucose-alanine cycle. The glucose-alanine cycle plays a role in recycling α-keto acids between the muscle and liver as well as is a mechanism for transporting amino nitrogen to the liver (the muscle cannot synthesize urea from amino nitrogen).

3.5. The pentose phosphate pathway

The pentose phosphate pathway is primarily a cytoplasmic anabolic pathway which converts the six carbons of glucose to five carbon sugars and reducing equivalents. Its principal products are NADPH + H^+, for reductive biosynthesis reactions within cells, and ribose-5-phosphate, for synthesis of the nucleotides and nucleic acids.

The pentose phosphate pathway occurs in the cytoplasm and is an alternative to glycolysis. There are two distinct phases in the pathway. The first is the oxidative phase. In this phase, two molecules of $NADP^+$ are reduced to NADPH + H^+, utilizing the energy from the conversion of glucose-6-phosphate into ribose-5-phosphate. The nonoxidative phase of the pathway primarily generates ribose-5-phosphate. This pathway also converts five carbon sugars into both six (fructose-6-phosphate) and three (glyceraldehyde-3-phosphate) carbon sugars which can then be utilized by the pathway of glycolysis.

4. Role of the pancreas in the glucose homeostasis

The pancreas plays a key role in the glucose homeostasis. The endocrine and exocrine pancreas has a complex anatomical and functional interaction [20]. Glucose metabolism is highly dependent on hormones secreted by the islets of Langerhans [21]. To avoid postprandial hyperglycemia and fasting hypoglycemia, the body can adjust glucose levels by secreting two hormones: insulin and glucagon. These hormones work in opposition to each other [22]. There are four major cell types in the pancreatic islets of Langerhans: the β-cells that secrete insulin and amylin, α-cells secrete glucagon, δ-cells secrete somatostatin, and PP cells secrete pancreatic polypeptide (PPY) [22, 23].

4.1. Insulin

Insulin secretion depends on the circulating glucose concentrations. It is secreted if the blood glucose concentration is ≥ 3 mmol/L [24]. Increased circulating glucose levels > 5 mmol/L results in increase of output of insulin and C peptide by the β-cells. Postprandially, the secretion of insulin occurs in two phases [26]. Long-term release of insulin occurs if glucose concentrations remain high [25]. Insulin secretion needs at least two signaling pathways, the K_{ATP} channel dependent and K_{ATP} channel independent, respectively [27, 28]. Glucose enters β-cells *via* GLUT2, which is believed to play a role in glucose-stimulated insulin secretion.

Insulin regulates glucose homeostasis at many sites, as for example, reducing hepatic glucose output (*via* decreased glucogenesis and glycogenolysis), inducing a process of glycogenesis (liver, muscle), and increasing the rate of glucose uptake, primarily into striated muscle and adipocytes. In most nonhepatic tissues, insulin increases glucose uptake by increasing the number of plasma membrane GLUT1 and GLUT4.

4.2. Glucagon

Glucagon is a hormone which is secreted by α-cells in response to hypoglycemia. It acts as the counter-regulatory hormone to insulin. Pancreatic α-cells contain a special set of channels that generate action potentials of Na^+ and Ca^{2+} in the absence or at low levels of glucose [29].

Glucagon activates glucose formation and release from the liver to stabilize blood glucose [30]. Glucagon stimulates gluconeogenesis and glycogenolysis and decreases glycogenesis and glycolysis. It also stimulates gluconeogenesis by stimulation of uptake of amino acids in the liver and increases the release of glycerol from adipose tissue which can further be used in the liver during gluconeogenesis [31]. An elevated glucagon-to-insulin ratio accelerates gluconeogenesis as well as fatty acid β-oxidation and ketone bodies formation [30, 32].

4.3. Somatostatin

Somatostatin is secreted by many tissues, including pancreatic δ-cells, intestinal tract, and central nervous system. It is released in response to glucose at lower concentrations than β-cells [33]. Somatostatin is a potent local inhibitor adjacent β- and α-cells [34]. Somatostatin release is increased in response to glucose stimulation [35, 36] and is Ca^{2+} dependent [36]. Acute administration of somatostatin to animals reduces food intake [37, 38]. Somatostatin has been reported to have no direct effect on basal glucose production (gluconeogenesis or glycogenesis) in isolated hepatocytes [39], and in vivo it does not alter the basal glucose production rate when the levels of insulin and glucagon are maintained [39, 40]. The portal vein insulin and glucagon levels were significantly decreased by somatostatin infusion [40].

4.4. Amylin

Amylin is produced by β-cells and stored in their secretory granules. It is co-secreted with insulin from pancreatic β-cells in response to glucose as same as insulin [41], in a roughly about 1:10 amylin/insulin ratio [42]. Plasma amylin levels are low during fasting and increase during meals and following glucose administration, and the levels are directly proportional to body fat [42]. Amylin participates in glucose homeostasis by two mechanisms: retarding gastric emptying in dose-response manner [43] and suppressing postprandial glucagon secretion [43, 44]. There is also evidence that amylin functions as an adiposity signal in addition to a satiety signal.

4.5. Pancreatic polypeptide (PPY)

The pancreatic polypeptide (PP) is produced predominantly by F cells (PP cells). Circulating PP concentrations increase following nutrient ingestion in a biphasic manner in proportion to the caloric load [45]. The secretion of PP during meals requires an intact vagus nerve. Pancreatic polypeptide affects metabolic functions including glycogenolysis and decreases fatty acid levels [46]. It also inhibits pancreatic secretion.

5. Role of the liver in the glucose homeostasis

The liver plays a major role in blood glucose homeostasis by maintaining a balance between the uptake and storage of glucose *via* glycogenolysis and gluconeogenesis. The liver is the primary organ for glucose metabolism. About 90% of all circulating glucose not derived directly from the diet comes from the liver. Hepatocytes take up glucose by GLUT2 in the presence of high concentrations of glucose. In hepatocytes, glucose is phosphorylated by glucokinase to glucose-6-phosphate. From glucose-6-phosphate, the glucose is directed into glycogenesis, the pentose phosphate pathway, or glycolysis.

In response to ingestion of glucose and the resulting hyperinsulinemia and hyperglycemia, the fasting liver shifts from net output to net uptake of glucose. Healthy human adults ingesting 75 g glucose exhibited peak plasma glucose and insulin concentrations of 7.8 mmol/L and 325 pmol/L, respectively [47]. Key enzymes in opposing metabolic pathways, glycolysis, and glycogenesis must be regulated for net flux in the appropriate direction to be achieved. The net glucose release is the result of two simultaneously ongoing pathways that are tightly regulated. Two enzymes specific for gluconeogenesis are opposed to the glycolytic enzymes. These enzymes regulate substrate cycles between gluconeogenesis and glycolysis. Glycogenolysis occurs within 2–6 hours after a meal in humans, and gluconeogenesis has a greater importance with prolonged fasting [48].

The rate of gluconeogenesis is controlled principally by the activation of gluconeogenic enzyme genes that are controlled by glucagon, glucocorticoids, and the interleukin-6 family of cytokines [48]. Insulin decreases gluconeogenesis by suppressing the expression of phosphoenolpyruvate carboxykinase and glucose-6-phosphatase, and glucagon and glucocorticoids stimulate glucose production by inducing these genes [49]. Glucagon is a regulator of hepatic glucose production during fasting, exercise, and hypoglycemia. It also plays a role in limiting hepatic glucose uptake. In response to a physiological rise in glucagon, hepatic glucose production is rapidly stimulated. This increase in hepatic glucose production is due to an enhancement of glycogenolysis, with little, or no, acute effect on gluconeogenesis [50]. The liver can release of glucose into the circulation. The skeletal muscle releases lactate, from where it can shuttle back to the liver (the Cori cycle).

The newborn mammals are in a transitional state of glucose homeostasis [51]. The diet of neonate is a low-carbohydrate, high-fat milk diet. The neonate must oxidize the stored liver glycogen, which is synthesized in the final days of gestation [51]. The initiation of hepatic glycogenolysis and gluconeogenesis in the first postnatal hours is critical for the maintenance of glucose homeostasis at this time [52]. Fetal life is characterized by chronic hyperinsulinemia. At birth hyperinsulinemia continues briefly and is one of the factors involved in the natural delay in hepatic glycogenolysis [53]. Counter-regulatory hormone actions are vital for the reversal of the postnatal hypoglycemia and for establishing glucose homeostasis at this time. Glucagon released in response to the postnatal hypoglycemia is responsible for initiation glycogenolysis and switching on hepatic gluconeogenesis [52].

6. Role of the kidneys in the glucose homeostasis

The human kidney is involved in the regulation of glucose homeostasis *via* three mechanisms: release of glucose into the circulation *via* gluconeogenesis, uptake of glucose from the circulation, and reabsorption of glucose from glomerular filtrate to conserve glucose carbon [54]. The kidney is unable to release glucose through glycogenolysis [55]. Glucose utilization occurs predominantly in the renal medulla. These enzymes can take up, phosphorylate, glycolyse, and accumulate, but cannot release, free glucose into the circulation. Glucose release is confined to the renal cortex [56]. Cells in the renal cortex possess gluconeogenic enzymes, and they can release glucose into circulation [57, 58]. The main precursor for renal glucogenesis is lactate [57]. Obtained results revealed that lactate is the most important renal gluconeogenic substrate followed by glutamine and glycerol [59]. Renal conversion to glucose of these precursors accounted for ~ 50, 70, and 35%, respectively, of their overall systemic gluconeogenesis [57]. After an overnight fast, 20–25% of glucose released into the circulation derives from the kidneys [54].

Renal glucogenesis is chiefly regulated by insulin and adrenaline. Insulin reduces renal gluconeogenesis and reduces the availability of gluconeogenic substrates, thus reducing glucose release into circulation [60]. On the other hand, insulin stimulates renal glucose uptake [61]. Adrenaline stimulates renal glucogenesis and glucose release and reduces renal glucose uptake [60]. It was shown in animal studies that glucagon increases renal glucose release into circulation.

With a daily glomerular filtration rate of 180 L, approximately 162 g of glucose must be reabsorbed each day to maintain a normal fasting plasma glucose concentration of 5.6 mmol/L [62]. Reabsorption of glucose in the proximal tubule is mediated by glucose transporter proteins that are present in cell membranes. SGLTs mediate active transport of glucose. SGLT2, which is in the convoluted section on the proximal tubule (S1), is considered most important. It is responsible for reabsorbing 90% of the glucose filtered at the glomerulus. SGLT1, which is found in the straight section of the proximal tubule (S3), contributes to the other 10% of glucose reabsorbed in the proximal tubule [63]. GLUT proteins are expressed at the basolateral membrane of the epithelial cells. These transporters release into circulation the glucose reabsorbed by SGLTs in the tubular cells. Glucose reabsorbed by SGLT2 is then released into the circulation *via* GLUT2 and reabsorbed by SGLT1 [64]. In the postabsorptive setting after an overnight fast, the kidneys utilize approximately 10% of all glucose utilized by the body. After meal ingestion, their glucose utilization increases in absolute sense [54].

7. Role of the hypothalamic-pituitary axis in the glucose homeostasis

The role of the brain to control glucose homeostasis was introduced in 1964 [65, 66]. Energy homeostasis is maintained by adapting meal size to current energy requirements. This control is achieved by communication between the digestive system and central nervous system. Two systems regulate the quantity of food intake: short term, which prevents overeating, and long term, involved in the energy stores as a fat [67]. Several regions of the brain are involved in regulation of food intake and energy homeostasis [68–72]. The hypothalamus is the most important locus involved in the neural control peripheral metabolism through the modulation of autonomic nervous system activity. The autonomic nervous system modulates hormone secretion (insulin and glucagon) and metabolic activity of the liver, adipose tissue, and muscle. The hypothalamus is in turn informed of the energy status of the organism. This is due to the metabolic and hormonal signals. There are two ways for the hypothalamus to signal to the peripheral organs: by stimulating the autonomic nerves and by releasing hormones from the pituitary gland. The hypothalamus consists of three areas: lateral, an important region regulating the cessation of feeding [73]; medial; and paraventricular, which is involved in the initiation of feeding [74]. In addition to direct neural connections, the hypothalamus can affect metabolic functions by neuroendocrine connections.

In the hypothalamus-pancreas axis, autonomic nerves release glucagon and insulin, which directly enter the liver and affect liver metabolism. In the hypothalamus-adrenal axis, autonomic nerves release catecholamines from adrenal medulla, which also affect liver metabolism. The hypothalamus-pituitary axis, which consists of neuroendocrine pathways from the hypothalamus, can also regulate liver functions. The hypothalamus sends signals to the pituitary gland, which release different hormones. Among them, three are thought to be intensely involved in the regulation of liver glucose metabolism [75].

The hypothalamic-pituitary-adrenal (HPA) axis referees to a complex set of homeostatic interactions between the hypothalamus, the pituitary gland, and the adrenal gland. The core of the HPA axis is the paraventricular nucleus (PVN) of the hypothalamus. The PVN contains neurocrine neurons, which synthesize and secrete vasopressin (AVP) and corticotrophin-releasing hormone (CRH). These two peptides can stimulate the secretion of the adrenocorticotropic hormone (ACTH) from anterior pituitary. In turn, ACTH enters peripheral circulation where it reaches the adrenal cortex to induce glucocorticoid hormone production (cortisol). Glucocorticoids exert a negative feedback on the paraventricular nucleus of the hypothalamus and pituitary to suppress CRH and ACTH production, respectively. Activation of glucocorticoids in vivo causes activation of glycogen synthase and inactivation of phosphorylase, resulting in glycogen synthesis [76]. Glucocorticoids lead to lipolysis in adipose tissue and proteolysis in the skeletal muscle by inhibiting glucose uptake by these tissues resulting in release of glycerol from adipose tissue and amino acids from the muscle [77, 78]. In turn, glycerol and amino acids are used as substrates to produce glucose in the liver. Glucocorticoids stimulate hepatic gluconeogenesis and antagonize actions of insulin in the liver and muscle, thus tending to increase glucose levels. The expression of GLUT4 is increased by glucocorticoids in the skeletal muscle and adipose tissue. Increased lipolysis may be important in glucocorticoid-induced insulin resistance. Glucocorticoids inhibit insulin secretion from pancreatic β-cells.

Maintenance of thyroid function is depended on a complex interplay between the hypothalamus, anterior pituitary, and thyroid gland (HPT). The thyroid gland is controlled by the activity of the hypothalamic-pituitary-thyroid axis. The hypothalamus releases thyrotropin-releasing hormone (TRH) which stimulates the biosynthesis, and release of thyrotropin (TSH) forms the anterior pituitary. TSH stimulates the thyroid gland which releases thyroxine (T4) and triiodothyronine (T3) into the circulation. Thyroid hormone action has been long recognized as a significant determinant of glucose homeostasis [79, 80]. Glucose homeostasis appears to be the result of the T3 and insulin synergistic regulation of gene transcription involved metabolic pathways of glucose and lipids [81]. T3 regulates a gene expression of glucose metabolism (the enzymes for oxidation of glucose and lipids, glucose storage, glycolysis, cholesterol synthesis, and glucose-lipid metabolism) [82]. T3 directly stimulates basal and insulin-mediated glucose uptake in the rat skeletal muscle. This induction was shown to be due primarily to an increase in Glut4 protein expression [83].

Human growth hormone (GH) is an essential regulator of carbohydrate and lipid metabolism. It increases indirectly the production of glucose in the liver. Glycerol released into the blood acts as a substrate for gluconeogenesis in the liver. GH antagonizes insulin action; increases fasting hepatic glucose output, by increasing hepatic gluconeogenesis and glycogenolysis; and decreases peripheral glucose utilization through the inhibition of glycogen synthesis and glucose oxidation [84].

The main regulatory factor of reproductive functions is gonadotropin-releasing hormone (GnRH), secreted by the hypothalamus. GnRH is a primary stimulator of luteinizing hormone (LH) and follicle-stimulating hormone (FSH). In men, LH stimulates testes to synthesis and secrete sex hormone, testosterone. In women, FSH acts on the ovary to stimulate and release estrogens. Estrogens are considered in blood glucose homeostasis. Estrogens have an adverse effect on carbohydrate metabolism. Administration of estrogens increases the insulin content of the pancreas in rats. In β-cells estrogens increase biosynthesis of proinsulin. During pregnancy, estrogen receptor integrates information from estrogen, glucose and other nutrients in the blood to regulate insulin gene expression and, therefore, contributes to the maintenance of insulin and glucose homeostasis [85]. Estrogen increases expression of glucose transporters and glucose transport in blood-brain barrier endothelium. Androgens can influence body composition, which is associated with insulin sensitivity. Testosterone may affect insulin sensitivity. Patients treated with androgen deprivation therapy have elevated glucose and increased insulin resistance. Testosterone treatment in hypogonadal men reduces fasting insulin. Testosterone activates the glucose metabolism-related signaling pathway in the skeletal muscle. The addition of testosterone to the cultured skeletal muscle induces the elevation of GLUT4 protein expression and accelerates its translocation from cytosol to plasma membrane. In women, testosterone induces selective insulin resistance in cultured subcutaneous adipocytes.

Author details

Leszek Szablewski

Address all correspondence to: leszek.szablewski@wum.edu.pl

Medical University of Warsaw, Warsaw, Poland

References

[1] Manel N, Kim FJ, Kinet S, Taylor N, Sitbon M, Battini JL. The ubiquitous glucose transporter GLUT-1 is a receptor for HTLV. Cell. 2003;**115**:449-459.

[2] Macintire AN, Gerriets VA, Nichols AG, Michalek RD, Rudolph MC, Deoliveira D, et al. The glucose transporter Glut1 is selectively essential for CD4 T cell activation and effector function. Cell Metab. 2014;**20**:61-72.

[3] Mueckler M, Thorens B. The *SLC2* (GLUT) family of membrane transporters. Mol Aspects Med. 2013;**34**:121-138.

[4] Augustin R. The protein family of glucose facilitators: it's not only about glucose after all. IUBMB Life. 2010;**62**:315-333.

[5] Zhao FQ, Keating AF. Functional properties and genomics of glucose transporters. Curr Genomics. 2007;**8**:113-128.

[6] Medina RA, Owen GI. Glucose transporters: expression, regulation and cancer. Biol Res. 2002;**35**:9-26.

[7] Wright EM. Glucose transport families *SLC5* and *SLC50*. Mol Aspects Med. 2013; **34**:183-196.

[8] Navale AM, Paranjape AN. Glucose transporters: physiological and pathological roles. Biophys Rev. 2016;**8**:5-9.

[9] Bianchi L, Diez-Sampedro A. A single amino acid change converts the sugar sensor SGLT3 into a sugar transporter. PLoS One. 2010;**5**:e10241.

[10] Wright EM, Loo DDF, Hirayama BA. Biology of human sodium glucose transporters. Physiol Rev. 2011;**91**:733-794.

[11] Wright EM. Renal Na^{+}/glucose cotransporters. Am J Physiol. 2001;**280**:F10–F18.

[12] Turk E, Wright EM. Membrane topology motifs in the SGLT cotransporters family. J Membr Biol. 1977;**159**:1-20.

[13] Drozdowski LA, Thomson ABR. Intestinal sugar transport. World J Gastroenterol. 2006;**12**:1657-1670.

[14] Wright EM. Glucose galactose malabsorption. Am J Physiol. 1998;**275**:G879–G882.

[15] Chen LQ, Hou BH, Lalonde S, Takanaga H, Hartung ML, Qu XQ, et al.. Sugar transporters for intracellular exchange and nutrition of pathogens. Nature. 2010;**468**:527-532.

[16] Feng L, Frommer WB. Structure and function of SemiSWEET and SWEET sugar transporters. Trends Biochem Sci. 2015;**40**:480-486.

[17] Tao Y, Cheung LS, Li S, Eom JS, Chen LQ, Xu Y, et al.. Structure of a eukaryotic SWEET transporter in a homo-trimeric complex. Nature. 2015;**527**:259-263.

[18] Loqué D, Lalonde S, Looger LL, von Wirén N, Frommer WB. A cytosolic trans-activation domain essential for ammonium uptake. Nature. 2007;**446**:195-198.

[19] Eom JS, Chen LQ, Sosso D, Julius BT, Lin IW, Qu XQ, et al.. SWEETs, transporters for intracellular and intercellular sugar translocation. Curr Opin Plant Biol. 2015;**25**:53-62.

[20] Pap A. Effects of insulin and glucose metabolism on pancreatic exocrine function. Int J Diabets Metab. 2004;**12**:30-34.

[21] Pendharkar SA, Asrani VM, Xiao AY, Yoon HD, Murphy R, Windsor JA, et al.. Relationship between pancreatic hormones and glucose metabolism: a cross-sectional study in patients after acute pancreatitis. Am J Physiol Gastrointest Liver Physiol. 2016;doi:10.1152/ajpgi.00074.2016.

[22] Szablewski L. Glucose homeostasis. In Glucose homeostasis and insulin resistance, Szablewski L. ed. Bentham eBooks, Sharjah, United Arab Emirates, 2011.

[23] Bermúdez-Silva FJ, Pérez JS, Nadal A, de Fonseca FR. The role of the pancreatic endocannabinoid system in glucose metabolism. Best Pract Res Clin Endocrinol Metab. 2009;**23**:87-102.

[24] Gerich JE. Control of glycemia. Bailliers Best Pract Res Clin Endocrinol Metab. 1993;**7**:551-586.

[25] Aronoff SL, Berkowitz K, Shreiner B, Want L. Glucose metabolism and regulation beyond insulin and glucagon. Diabetes Spectr. 2004;**17**:183-190.

[26] Henquin JC, Ishiyama N, Nenquin M, Ravier MA, Jonas JC. Signals and pools underlying biphasic insulin secretion. Diabetes. 2002;**51**:S60–S67.

[27] Straub SG, Sharp GWG. Glucose-stimulated signaling pathways in biphasic insulin secretion. Diabetes Metab Res Rev. 2002;**18**:451-463.

[28] Rorsman P. Insulin secretion: function and therapy of pancreatic beta-cells in diabetes. Br J Diabetes Vasc Dis. 2005;**5**:187-191.

[29] Rorsman P, Salehi SA, Abdulkader F, Braun M, MacDonald PE. K_{ATP}-channels and glucose regulated glucagon secretion. Trends Endocrinol Metab. 2008;**19**:277-284.

[30] Quesada J, Tuduri E, Ripoll C, Nadal A. Physiology of the pancreatic α-cell and glucagon secretion: role in glucose homeostasis and diabetes. J Endocrinol. 2008;**199**:5-19.

[31] Gromada J, Franklin I, Wollheim C. Alpha-cells of the endocrine pancreas: 35 years of research but enigma remains. Endocr Rev. 2007;**28**:84-116.

[32] Vons C, Pegorier JP, Giard J, Kohl C, Ivanov MA, Franco D. Regulation of fatty-acid metabolism by pancreatic hormones in cultured human hepatocytes. Hepatology. 1991;**13**:1126-1130.

[33] Nadal A, Quesada I, Soria B. Homologous and heterologous asynchronicity between identified alpha-, beta- and delta-cells within intact islets of Langerhans in the mouse. J Physiol. 1999;**517**:85-93.

[34] Stangner JL, Samols E. The vascular order of islet cellular perfusion in the human pancreas. Diabetes. 1992;**41**:93-97.

[35] Hermansen K, Christensen SE, Orskov H. Characterization of somatostatin release from the pancreas: the role of potassium. Scand J Clin Lab Invest. 1979;**39**:717-722.

[36] Kanno T, Göpel SO, Rorsman P, Wakui M. Cellular function in multicellular system for hormone secretion: electrophysiological aspect of studies on α- β- and δ-cells in the pancreatic islet. Neurosci Res. 2002;**42**:79-90.

[37] Levine AS, Morley JE. Peripheral administered somatostatin reduces feeding by the vagal mediated mechanism. Pharmacol Biochem Behav. 1982;**16**:897-902.

[38] Lotter EC, Krinsky R, McKay JM, Treneer CM, Porte D Jr, Woods SC. Somatostatin decreases food intake in rats and baboons. J Comp Physiol Psychol. 1981;**95**:278-287.

[39] Cherrington AD, Caldwell MD, Diets MR, Exton JH, Crofford DB. The effects of somatostatin on glucose uptake and production by rat tissues in vitro. Diabetes. 1977;**26**:740-748.

[40] Ogihara M, Ui M. Effects of somatostatin on liver glycogen and fat metabolism in vivo. Jpn J Pharmacol. 1984;**34**:313-318.

[41] Cooper GLS, Willis AC, Clark A, Turner RS, Sim RB, Reid KBM. Purification and characterization of a peptide from amyloid-rich pancreas of the type 2 diabetic patients. Proc Natl Acad Sci U S A. 1987;**84**:8628-8632.

[42] Woods SC, Lutz TA, Geary N, Langhans W. Pancreatic signals controlling food intake; insulin, glucagon and amylin. Philos Trans R Soc B Biol Sci. 2006;**361**:1219-1235.

[43] Lutz TA. Amylinergic control of food intake. Physiol Behav. 2006;**89**:465-471.

[44] Gedulin BR, Rink TJ, Young AA. Dose-response for glucagonostic effect of amylin in rats. Metabolism. 1997;**46**:67-70.

[45] Track NS, McLeod RS, Mee AV. Human pancreatic polypeptide studies of fasting and postprandial plasma concentrations. Can J Physiol Pharmacol. 1980;**58**:1484-1489.

[46] Gehlert DR. Multiple receptors for the pancreatic polypeptide (PP-fold) family: physiological implications. Proc Soc Exp Biol Med. 1998;**218**:7-22.

[47] Meyer C, Dostou JM, Welle SL, Gerich JE. Role of human liver, kidney, and skeletal muscle in postprandial glucose homeostasis. Am J Physiol Endocrinol Metab. 2002;**282**:E419–E427.

[48] Postic C, Dentin R, Girard J. Role of the liver in the control of carbohydrate and lipid metabolism. Diabetes Metab. 2005;**30**:398-408.

[49] O'Brien RM, Granner DK. Regulation of gene expression by insulin. Physiol Rev. 1996;**76**:1109-1161.

[50] Ramnanan CJ, Edgerton DS, Kraft G, Cherrington AD. Physiologic action of glucagon on liver glucose metabolism. Diabetes Obes Metab. 2011;**13**(Suppl 1):118-125.

[51] Snell K. Regulation of hepatic glucose metabolism by insulin and counter-regulatory hormones. Proc Nutr Soc. 1991;**50**:567-575.

[52] Girard JR, Cuendet GS, Marliss EB, Kervran A, Rieutort M, Assan R. Fuels, hormones, and liver metabolism at term and during the early postnatal period in the rat. J Clin Invest. 1973;**52**:3190-3200.

[53] Snell K, Walker DG. Glucose metabolism in the newborn rat: temporal studies in vivo. Biochem J. 1973;**132**:739-752.

[54] Gerich JE. Role of the kidney in normal glucose homeostasis and in hyperglycaemia of diabetes mellitus: therapeutic implications. Diabet Med. 2010;**27**:136-142.

[55] Stumvoll M, Meyer C, Mitrakou A, Nadkarni V, Gerich JE. Renal glucose production and utilization: new aspects in humans. Diabetologia. 1997;**40**:749-757.

[56] Schoolwerth A, Smith B, Culpepper R. Renal gluconeogenesis. Miner Electrolyte Metab. 1988;**14**:347-361.

[57] Gerich JE, Meyer C, Waerle HJ, Stumvoll M. Renal gluconeogenesis: its importance in human homeostasis. Diabetes Care. 2001;**24**:382-391.

[58] Cano N. Bench-to-bedside review: glucose production from the kidney. Crit Care. 2002;**6**:317-321.

[59] Meyer C, Stumvoll M, Welle S, Kreider M, Nair S, Gerich J. Human kidney substrate utilization and gluconeogenesis. Diabetologia. 1974;**40**(Suppl 1):A24.

[60] Wilding JPH. The role of the kidneys in glucose homeostasis in type 2 diabetes: clinical implications and therapeutic significance through sodium glucose co-transporter 2 inhibitors. Metabolism. 2014;**63**:1228-1237.

[61] Cersosimo E, Judd R, Miles J. Insulin regulation of renal glucose metabolism in conscious dogs. J Clin Invest. 1994;**93**:2584-2589.

[62] Poudel RR. Renal glucose handling in diabetes and sodium glucose cotransporter 2 inhibition. Indian J Endocrinol Metab. 2013;**17**:588-593.

[63] Wright EM, Hirayama BA, Loo DF. Active sugar transport in health and disease. J Intern Med. 2007;**261**:32-43.

[64] Wood IS, Trayhurn P. Glucose transporters (GLUT and SGLT): expanded families of sugar transport protein. Br J Nutr. 2003;**89**:3-9.

[65] Anand BK, Chhina GS, Sharma KN, Dua S, Singh B. Activity of single neurons in the hypothalamic feeding centers: effect of glucose. Am J Physiol. 1964;**207**:1146-1154.

[66] Oomura Y, Kimura K, Ooyama H, Maeno T, Iki M, Kuniyoshi N. Reciprocal activities of the ventromedical and lateral hypothalamic area of cats. Science. 1964;**143**:484-485.

[67] Konturek SJ, Konturek PC, Konturek JW, Cześnikiewicz-Guzik M, Brzozowski T, Sito E. Neuro-hormonal control of food intake; basic mechanisms and clinical implications. J Physiol Pharmacol. 2005;**56**(Suppl 6):5-25.

[68] Lyngdoh JA, Marbaniang E, Lynrah KG, Lyngdoh M. The role of brain in regulation of glucose homeostasis. Int J Med Sci Publ Health. 2015;**4**:1477-1480.

[69] Scarlett JM, Schwartz MW. Gut-brain mechanism controlling glucose homeostasis. F1000Prime Reports. 2015;**7**:12

[70] Rogoff D, Ryder JW, Black K, Yan Z, Burgess SC, McMillan DR, et al.. Abnormalities of glucose homeostasis and the hypothalamic-pituitary-adrenal axis in mice lacking hexose-6-phosphate dehydrogenase. Endocrinology. 2007;**148**:5072-5080.

[71] Si MW, Yang MK, Fu XD. Effect of hypothalamic-pituitary axis alterations on glucose and lipid metabolism in diabetic rats. Genet Mol Res. 2015;**14**:9562-9570.

[72] Perry RJ, Zhang XM, Zhang D, Kumashiro N, Camporez JPG, Cline GW, et al.. Leptin reverses diabetes by suppression of the hypothalamic-pituitary-adrenal axis. Nat Med. 2014;**20**:759-763.

[73] Corbett SW, Kaufman LN, Keesey RE. Thermogenesis after lateral hypothalamic lesions: contributions of brown adipose tissue. Am J Physiol. 1988;**255**:E708–E715.

[74] Williams G, Bing C, Cai XJ, Harrold JA, King PJ, Liu XH. The hypothalamus and the control of energy homeostasis: different circuits, different purposes. Physiol Behav. 2001;**74**:683-701.

[75] Uyama N, Geerts A, Reynaert H. Neural connections between the hypothalamus and the liver. Anat Rec Discov Mol Cell Evol Biol. 2004;**280**:808-820.

[76] Laloux M, Stalmans W, Hers HG. On the mechanism by which glucocorticoids cause the activation of glycogen synthase in mouse and rat livers. Eur J Biochem. 1983;**136**:175-181.

[77] Livingston JN, Lockwood DH. Effect of glucocorticoids on the glucose transporter system in isolated fat cells. J Biol Chem. 1975;**250**:8353-8360.

[78] Smith OL, Wong CY, Gelfand RA. Influence of glucocorticoids on skeletal muscle proteolysis in normal and diabetic-adrenolectomized eviscerated rats. Metabolism. 1990;**39**:641-646.

[79] Muller MJ, Seite HJ. Interaction of thyroid hormones and cyclic AMP in the stimulation of hepatic gluconeogenesis. Biochem Biophys Acta. 1983;**756**:360-368.

[80] Nebioglu S, Wathanaronchai P, Nebioglu D, Pruden EL, Gibson DM. Mechanisms underlying enhanced glycogenolysis in livers of 3,5,3′-triiodothyronine-treated rats. Am J Physiol. 1990;**258**:E109–E116.

[81] Granner DK, Pilkis S. The genes of hepatic glucose metabolism. J Biol Chem. 1990;**265**:10173-10182.

[82] Kim SR, Talbott EA, Tull E, Vogt M, Andersen S, Kuller LH. Contribution of abnormalities of thyroid hormones to type 2 diabetes. Diabetes Care. 2000;**23**:260-261.

[83] Casla A, Rovira A, Wells JA, Dohm GL. Increased glucose transporter (GLUT4) protein expression in hyperthyroidism. Biochem Biophys Res Commun. 1990;**171**:182-188.

[84] Fowelin J, Attval S, von Schenck H, Smith U, Lager I. Characterization of the insulin-antagonistic effect of growth hormone in man. Diabetologia. 1991;**34**:500-506.

[85] Nadal A, Alonso-Magdalena P, Soriano S, Quesada I, Ropero AB. The pancreatic β-cell as a target of estrogens and xenoestrogens: implications for blood glucose homeostasis and diabetes. Mol Cell Endocrinol. 2009;**304**:63-68.

Chapter 3

Gluconeogenesis: A Metabolic Pathway in Eukaryotic Cells such as Cellular Slime Molds

Richa Karmakar

Additional information is available at the end of the chapter

http://dx.doi.org/10.5772/67221

Abstract

Dictyostelium discoideum or cellular slime mold is simple eukaryotic microorganism, which generally grows in forest soil and decaying leaves. This amoeba feeds on bacteria and grows as single cells. The development of *Dictyostelium discoideum* is simpler than that of mammalian cells. It uses many of the same signals that are found to function in higher eukaryotic organisms like plants and animals. *Dictyostelium discoideum* is an excellent system in which to study metabolic pathways which are simpler than that of the complex systems like mammalian system. Glucose is metabolized in glycolysis to yield pyruvate and lactate and further metabolized in the tricarboxylic acid cycle. Glucose can be polymerized into glycogen in addition to glycolysis process. In a metabolic pathway, the generation of glucose from certain non-carbohydrate carbon substrates is called gluconeogenesis. In *Dictyostelium discoideum*, glucose is synthesized by the breakdown of pyruvate. Glycogen phosphorylase and amylase break down glycogen to form glucose. Glycogen synthase and glycogen phosphorylase are the key enzymes for the regulation. Both the enzyme equally regulated the process simultaneously, so that when one is activated, the other is deactivated. During gluconeogenesis, glucose is synthesized from pyruvate but sometimes during this process, three enzymes, glucose-6-phophatase, fructose-1,6-bisphosphatase, and phosphoenolpyruvate carboxykinase catalyze an irreversible reaction.

Keywords: gluconeogenesis, eukaryotic system, *Dictyostelium discoideum*

1. Introduction

An amoeba is very interesting organism to study because it grows as single cells and develops as multi-cellular organisms. They present a range of developmental processes which can be used to study of any molecular pathways like glycolysis or gluconeogenesis pathways [1]. During evolution, the amoebozoa generated a large number of species which goes through the similar developmental stages, from unicellular to multi-cellular stages [1–4]. The well-characterized amoebozoan species is *Dictyostelium discoideum,* which is easy to study as compared to mammalian cell [5–7]. *D. discoideum* uses many similar signals and contains similar pathways that are presented in plants and animals [1]. *D. discoideum* or cellular slime mold is simple eukaryotic microorganism which generally grows in forest soil and decaying leaves. This amoeba feeds on bacteria and grows as single cells [2, 3]. The development of *D. discoideum* is simpler than that of mammalian cells.

2. The gluconeogenesis process in eukaryotic cell

Gluconeogenesis is a process by which carbohydrate is synthesized from non-carbohydrate precursors like oxaloacetate and pyruvate (**Figure 1**). In the first step of the gluconeogenesis process, oxaloacetic acid is synthesized from pyruvic acid. On the other hand, in the citric acid cycle, oxaloacetic acid reacts with acetyl-CoA. So, at low concentration of acetyl-CoA and high concentration of ATP, gluconeogenesis proceeds. Gluconeogenesis starts in the mitochondria of the cells. In the first step, carboxylation of pyruvate occurs by pyruvate carboxylase enzyme and it forms oxaloacetate by using one ATP molecule. Oxaloacetate is reduced to malate by using NADH. After this step, the remaining steps of gluconeogenesis process occur in the cytosol. In the next step, malate is oxidized to oxaloacetate using NAD^+. Oxaloacetate is first decarboxylated, and after that, it is phosphorylated by using the enzyme, PEP carboxykinase, and one GTP. In the next step, PEP converted into 2-phosphoglycerate, 3-phosphoglycerate and then 1,3-bisphosphoglycerate by the enzyme enolase, phosphoglycerate mutase and phosphoglycerate kinase, respectively. In the next step of this reaction, 1,3-bisphosphoglycerate converts into glyceraldehyde 3-phosphate by the enzyme glyceraldehyde phosphate dehydrogenase. Now, the glyceraldehyde 3-phosphate converts into fructose 1,6-bisphosphate via two ways: one is direct conversion and another through the intermediate component called dihydroxyacetone phosphate. In the next step, fructose 1,6-bisphosphate converts into fructose 6-phosphate, using an enzyme, fructose 1,6-bisphosphatase, one water molecule, and releasing one phosphate. This step is the rate-limiting step in gluconeogenesis process. Glucose-6-phosphate is formed from fructose 6-phosphate followed by glucose by the enzyme glucose-6-bisphosphatase. The reaction of the glucose formation occurs inside the endoplasmic reticulum, specifically in the lumen, where glucose-6-phosphate is hydrolyzed and produces glucose and releases an inorganic phosphate [8].

http://dx.doi.org/10.5772/67221

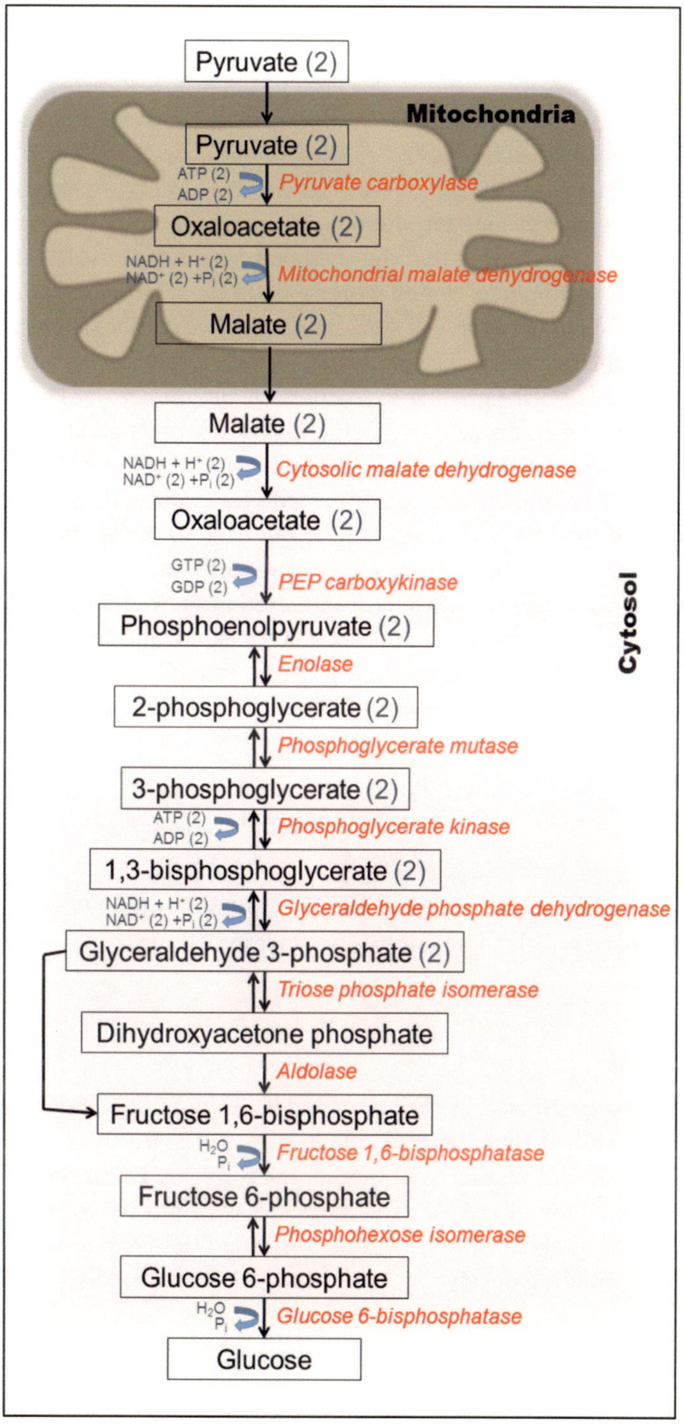

Figure 1. Gluconeogenesis process in the eukaryotic cell.

3.The developmental stages of *Dictyostelium discoideum.*

About 80 years ago, Ken Raper isolated *D. discoideum* from the forest floor at North Carolina [9]. He observed that when the cells had a depletion of food, they aggregated into mounds [9]. John Bonner showed that when the cells were in the starving condition, they secreted a chemical which acts like chemoattractant and the cells responded by moving up the gradient [10]. After 20 years, Konijn et al. showed that the chemoattractant was cAMP [11]. After this discovery, *D. discoideum* considered as a model organism to study chemotaxis and developmental biology. The connection between cell signaling pathways and biochemical pathways was established by using this organism. The pathways for cAMP synthesize, the surface receptors for cAMP and many other cell signaling to biochemical and molecular biological techniques were established by using *D. discoideum* [12–15]. The developmental stages of *D. discoideum* started from slug-shaped structures and go till the formation of the fruiting body [1, 2, 16]. Pre-spore and pre-stalk cells at the slug stage formed spores and stalk cells in fruiting bodies, and it was also found that pre-stalk cells were at the front of the slugs, and pre-spore cells were all in the back [2, 16]. There were 20-fold differences between the size of slugs and the total number of individual cells in each slug. This variation showed that there was some intracellular signal which determines the proportions of pre-spore and pre-stalk cells. Soderbom and Loomis showed that the pre-spore cells synthesize an inhibitor which inhibits pre-spore differentiation, and the pre-spore cells were resistant to this inhibitor [2, 17, 18]. This mechanism was responsible for size invariance of the slugs [18]. The amoeba went through three cycles when it was facing starvation [7].

3.1. The microcyst

Encystment was a very common process to amoebae, but it was not known for *D. discoideum*. In microcyst stage, each cell elaborated into a two layered cellulose coat and went to the dormant stage [7].

3.2. The macrocyst

In this stage of the sexual cycle, cells of two mating types fused [2]. Under wet condition, the macrocyst form, which had three layered cellulose coat at maturity. After fusion, the cells formed giant cells which had at least two nuclei or many nuclei. This fused structure attracted other amoebae by chemotaxis to cAMP. The endocytes were formed by engulfing these cells, and after that the giant cells produced meiotic offspring [7]. Macrocysts were formed from endocytes including hundred of cells.

3.3. Fruiting bodies

The fruiting body was formed through complex and polarized cell movements. In this stage, cells were not engulfed to form endocytes because one cell was recognized by the other cell. For the formation of fruiting body of *D. discoideum,* cells did chemotaxis and cells were more elaborated and involved a relay mechanism. But this mechanism either suppressed or did not

http://dx.doi.org/10.5772/67221

exist during the macrocyst formation. Fruiting body formed by aggregation of one lakh cells. In this case, cells were adhesive in nature and moved among each other and able to distinguish between cAMP and other molecules [7].

During mid-developmental stage, *D. discoideum* can choose between two different pathways, one is from the finger stage, it can directly precede to culmination, or it can fall over to form a phototactic, migratory slug. This migratory stage is important for cells to find an appropriate site for fruiting body formation [19]. Slugs prefer dark and low ionic strength environment [20]. Two cell types were there inside the slug which implied that they were connected with the signaling system [7]. Twenty percentage cells died for the formation of the stalk of the fruiting body, and eighty percentage cells survive by the formation of spores. **Figure 2** represents the developmental stages of *D. discoideum*.

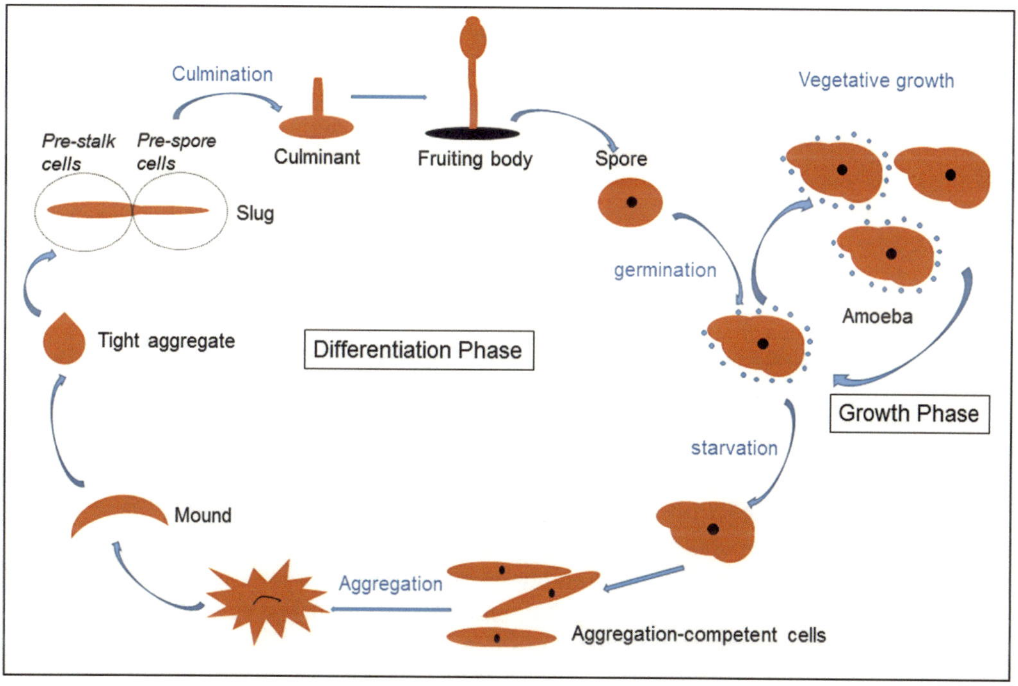

Figure 2. Lifecycle of *Dictyostelium discoideum*.

4. Size of the aggregates of *D. discoideum* is depending upon the gluconeogenesis pathway indirectly

The group size of *D. discoideum* is regulated by a negative feedback pathway mediated by counting factor (CF) which consists of at least five protein complexes [24]. During the early developmental stage, the counting factor (CF) breaking up the big aggregate of cells to the smaller aggregates of about 2×10^4 cells [21]. High levels of CF decrease cell-cell adhesion and also decrease the amplitude of cAMP and increase random motility. The glucose metabolism is affected by the CF, and it decreases the CF glucose levels [22]. CF decreases the activity of the gluconeogenic enzyme, glucose-6-phosphatase, which decreases the level of glucose in the cell with the high secretion of CF [23, 24]. In that case, if glucose has been added externally then the size of the fruiting bodies get increases [24]. This process alters the intermediates of the metabolic pathways, such as pyruvate and lactate [22]. Jang and Gomer showed that, if the cells exposed to CF, the CF has very small effect on amylase or glycogen phosphorylase, enzymes involved in glucose production from glycogen [24]. On the other hand, it has a huge effect on glycolysis pathway. If the CF is high, then it is inhibited the glucokinase activity, but it does not regulate phosphofructokinase (enzyme responsible for glycolysis pathways). CF showed some effect upon the enzyme involved in a gluconeogenesis pathway such as fructose-1,6-bisphosphatase and glucose-6-phosphatase. The fructose-1,6-bisphosphatase is not regulated by CF, whereas glucose-6-phosphatase is regulated by CF [24].

The size of the terminal structures of *D. discoideum* is depending upon the cell surrounded by the number of cells, so the initiation of development is not started with the significant growth of the cell. The secretion of a protein complex is very important to control the size of the aggregates. Large aggregates initially have higher CF levels which can modify gluconeogenesis, and the other metabolic pathways alters the level of metabolites [1, 24]. After that, the level of CF drops down which increases cell-cell adhesion and because of that the random motility decreases which stabilize the smaller aggregates.

5. Gluconeogenesis process affected during the differentiation of myxamoebae of the cellular slime mold

Myxamoeba is a naked amoeboid uni-nucleate protoplast that lacks both cilia and flagella. In the life cycle of *D. discoideum,* they had gone through the vegetative state, differentiation state as independent amoeboid cells which called as myxamoebae [25]. If the myxamoebae was grown in different media, with varying carbohydrate content [26], then the chemical composition [27], enzyme composition [28], and physiological behavior [29] got changed inside the cell. After that, if these cells put in the moist condition, the cells formed slug [5, 25]. Carbohydrate content changed inside the *D. discoideum* at different stages of development [30]. So the gluconeogenesis process alters during the carbohydrate conversion process, the energy was coming from cellular protein, RNA, and dry weight to complete this process [25, 30–33].

Wright et al. showed in their kinetic model, the glycogen content of the cell remained constant during the early developmental stage but it decreased when the culmination process occurred [34]. White and Sussman suggested that the glycogen content of axenic cells was small [30]. They also showed that the glycogen initially decreased because of the consumption of the bacterial glycogen for the development of axenically grown myxamoebae. Wright et al. assume in their model, during differentiation of *D. discoideum,* there was no gluconeogenesis process

http://dx.doi.org/10.5772/67221

occurs [34]. Cleland and Coe showed some evidence of the presence of gluconeogenesis process during the differentiation of myxamoebae [31]. However, Hames et al. suggested that initially, cells had low glycogen but during differentiation, the glycogen content increased significantly which depicts that the gluconeogenesis process had a huge effect on the differentiation mechanism of the cell [25, 35].

In the absence of glucose and the presence of very low concentrations of glycogen, myxamoebae grow and degrade all the glycogen during 4 h of development [35]. Glycogen is synthesized during the late developmental stages (5–15 h) and finally broken down by the cell to synthesize saccharide. Hames and Ashworth showed that the amount of glycogen synthesized during the late developmental stage is larger than the glycogen content of the vegetative cells [25]. This glycogen synthesis occurs during gluconeogenesis process when the cellular glucose remains at a constant low concentration. During differentiation, myxamoebal glycogen is not stored, but the gluconeogenesis process still can occur, if the cells initially have a large amount of glycogen.

6. Discussion

D. discoideum is a well-characterized eukaryotic system which grows as single cells and develops as multi-cellular organisms. The development of *D. discoideum* is simpler than that of mammalian cells. They use many similar signals and contain similar pathways that are presented in plants and animals so it is the best biological system to study the molecular pathways like glycolysis or gluconeogenesis pathways, which can correlated with the mammalian system. *D. discoideum* considered as a model organism to study chemotaxis and developmental biology. *D. discoideum* obtained energy through lysosomal degradation and phagocytosis process [36]. Because of that intra-lysosomal nutrient levels are increasing so, the lysosome and the vacuole collect amino acids and other nutrients from cellular components through the autophagy process [37]. This process has been correlated with the gluconeogenesis process. The counting factor (CF) or the protein complex regulates the size of the aggregates of *D. discoideum*, which is indirectly affecting the gluconeogenesis pathway. CF showed some effect upon the enzyme involved in a gluconeogenesis pathway such that fructose-1,6-bisphosphatase and glucose-6-phosphatase. The fructose-1,6-bisphosphatase is not regulated by CF, whereas glucose-6-phosphatase is regulated by CF, which acts as one of the size regulating factor for the aggregates of *D. discoideum*. Large aggregates initially have higher CF levels which can modify gluconeogenesis, and the other metabolic pathways, and also alters the level of metabolites [1, 24]. After that, the level of CF drops down which increases cell-cell adhesion and because of that the random motility decreases which stabilize the smaller aggregates. The glycogen synthesis during the developmental process of *D. discoideum* causes by the gluconeogenesis pathway. During differentiation, myxamoebal glycogen is not stored but the gluconeogenesis process still can occur, if the cells initially have a large amount of glycogen. So gluconeogenesis pathway is very important for the development of the cells, this is one of the best eukaryotic system in which the process can be studied.

Author details

Richa Karmakar

Address all correspondence to: richa1503@gmail.com

Department of Physics, University of California San Diego, La Jolla, CA, USA

References

[1] Loomis, W. F. (2014). Cell signaling during development of *Dictyostelium.* Dev. Biol.; 391:1-16.

[2] Raper, K. B. (1984). The Dictyostelids. Princeton University Press, Princeton, NJ.

[3] Eichinger, L., Pachebat, J. A., Glockner, G., Rajandream, M. A., Sucgang, R., Berriman, R., Song, J., Olsen, R., Szafranski, K., Xu, Q., Tunggal, B., Kummerfeld, S., Madera, M., Konfortov, B. A., Rivero, F., Bankier, A. T., et al. (2005). The genome of the social amoeba *Dictyostelium discoideum.* Nature; 435:43-57.

[4] Schilde, C., Schaap, P. (2013). The amoebozoa. Methods Mol. Biol.; 983:1-15.

[5] Bonner, J. T. (1959). The Cellular Slime Molds. Princeton University Press, Princeton, NJ.

[6] Loomis, W. F. (1975). *Dictyostelium discoideum*—A Developmental System. Academic Press, New York.

[7] Kessin, R. H. (2001). *Dictyostelium*—Evolution, Cell Biology, and the Development of Multicellularity. Cambridge University Press, Cambridge, UK.

[8] Nelson, D. L., Cox, M. M. (2000). Lehninger Principles of Biochemistry. Worth Publishers, USA.

[9] Raper, K. B. (1935). *Dictyostelium discoideum,* a new species of slime mold from decaying forest leaves. J. Agric. Res.; 50:135-147.

[10] Bonner, J. T. (1947). Evidence for the formation of cell aggregates by chemotaxis in the development of the slime mold *Dictyostelium discoideum.* J. Exp. Zool.; 106:1-26.

[11] Konijn, T. M., VandeMeene, J. G. C., Bonner, J. T., Barkley, D. S. (1967). The acrasin activity of adenosine-3′,5′-cyclicphosphate. Proc. Natl. Acad. Sci. USA; 58:1152-1154.

[12] Klein, P., Vaughan, R., Borleis, J., Devreotes, P. (1987). The surface cAMP receptorin Dictyostelium-levels of ligand-induced phosphorylation, solubilization, identification of primary transcript, and developmental regulation of expression. J. Biol. Chem.; 262:358-364.

[13] Insall, R., Kuspa, A., Lilly, P. J., Shaulsky, G., Levin, L. R., Loomis, W. F., Devreotes, P. (1994). CRAC, a cytosolic protein containing a pleckstrin homology domain, is required for receptor and G protein-mediated activation of adenylyl cyclase in *Dictyostelium.* J. Cell Biol.; 126:1537-1545.

[14] Maeda, M., Aubry, L., Insall, R., Gaskins, C., Devreotes, P. N., Firtel, R. A. (1996). Seven helix chemoattractant receptors transiently stimulate mitogen-activated protein kinase in *Dictyostelium*—role of heterotrimeric G proteins. J. Biol. Chem.; 271:3351-3354.

[15] Swaney, K. F., Huang, C. H., Devreotes, P. N. (2010). Eukaryotic chemotaxis: a network of signaling pathways controls motility, directional sensing, and polarity. Annu. Rev. Biophys.; 39:265-289.

[16] Raper, K. B. (1940). Pseudoplasmodium formation and organization in *Dictyostelium discoideum*. J. Elisha Mitchell Sci. Soc.; 56:241-282.

[17] Loomis, W. F. (1993). Lateral inhibition and pattern formation in *Dictyostelium*. Curr. Top. Dev. Biol.; 28:1-46.

[18] Soderbom, F., Loomis, W. F. (1998). Cell-cell signaling during *Dictyostelium* development. Trends Microbiol.; 6:402-406.

[19] Bonner, J. T., Davidowski, T. A., Hsu, W. L., Lapeyrolerie, D. A., Suthers, H. L. B. (1982). The role of surface water and light on differentiation in the cellular slime molds. Differentiation; 21:123-126.

[20] Sussman, M. (1987). Cultivation and synchronous morphogenesis of *Dictyostelium* under controlled experimental conditions. Methods Cell. Biol.; 28:9-29.

[21] Tang, L., Gao, T., McCollum, C., Jang, W., Vicker, M. G., Ammann, R. R., Gomer, R. H. (2002). A cell number-counting factor regulates the cytoskeleton and cell motility in *Dictyostelium*. Proc. Natl. Acad. Sci. USA.; 99:1371-1376.

[22] Jang, W., Schwartz, O. G., Gomer, R. H. (2009). A cell number counting factor alters cell metabolism. Commun. Integr. Biol.; 2:293-297.

[23] Jang, W., Chiem, B., Gomer, R. H. (2002). A secreted cell number counting factor represses intracellular glucose levels to regulate group size in *Dictyostelium*. J. Biol. Chem.; 277:39202-39208.

[24] Jang, W., Gomer, R. H. (2005). Exposure of cells to a cell number-counting factor decreases the activity of glucose-6-phosphatase to decrease intracellular glucose levels in *Dictyostelium discoideum*. Eukaryot. Cell.; 4:72-81.

[25] Hames, B. D., Ashworth, J. M. (1974). The metabolism of macromolecules during the differentiation of myxamoebae of the cellular slime mould *Dictyostelium discoideum*. Biochem. J.; 142:301-315.

[26] Watts, D. J., Ashworth, J. M. (1970). Growth of myxamoebae of the cellular slime mould *Dictyostelium discoideum* in axenic culture. Biochem. J.; 119:171-174.

[27] Weeks, G., Ashworth, J. M. (1972). Glycogen Synthetase and the control of glycogen synthesis in the cellular slime mould *Dictyostelium discoideum* during the growth (Myxamoebal) phase. Biochem. J.; 126:617-626.

[28] Ashworth, J. M., Quance, J. (1972). Enzyme synthesis in myxamoebae of the cellular slime mould *Dictyostelium discoideum* during growth in axenic culture. Biochem. J.; 126:601-608.

[29] Ashworth, J. M., Watts, D. J. (1970). Metabolism of the cellular slime mould *Dictyostelium discoideum* grown in axenic culture. Biochem. J.; 119:175-182.

[30] White, G. J., Sussman, M. (1961). Metabolism of major cell components during slime mold morphogenesis. Biochim. Biophys. Acta.; 53:285-293.

[31] Cleland, S. V., Coe, E. L. (1969). Conversion of aspartic acid to glucose during culmination in *Dictyostelium discoideum*. Biochim. Biophys. Acta; 192:446-454.

[32] Gregg, J. H., Bronsweig, R. D. (1956). Biochemical events accompanying stalk formation in the slime mold, *Dictyostelium discoideum*. J. Cell. Comp. Physiol.; 48:293-299.

[33] Wright, B. E., Anderson, M. L. (1960). Protein and amino acid turnover during differentiation in the slime mold, I. Utilization of endogenous amino acids and proteins. Biochim. Biophys. Acta.; 43:62-66.

[34] Wright, B. E., Simon, W., Walsh, B. T. (1968). A kinetic model of metabolism essential to differentiation in *Dictyostelium discoideum*. Proc. Nat. Acad. Sci. USA.; 60:644-651.

[35] Hames, B. D., Weeks, G., Ashworth, J. M. (1972). Glycogen synthetase and the control of glycogen synthesis in the cellular slime mould *Dictyostelium discoideum* during cell differentiation. Biochem. J.; 126:627-633.

[36] Neuhaus, E. M., Almers, W., Soldati, T. (2002). Morphology and dynamics of the endocytic pathway in *Dictyostelium discoideum*. Mol. Biol. Cell.; 13:1390-1407.

[37] Efeyan, A., Comb, W. C., Sabatini, D. M. (2015). Nutrient-sensing mechanisms and pathways. Nature; 517:302-310.

Chapter 4

Neuroendocrine Control of Hepatic Gluconeogenesis

Zhuo Mao and Weizhen Zhang

Additional information is available at the end of the chapter

http://dx.doi.org/10.5772/67535

Abstract

Glucose is intricately regulated in human body through a complex network of hormonal and neuronal factors. A series of evidence suggests that the gastrointestinal tract and the central nervous system play prominent roles in the regulation of glucose and energy homeostasis. The gut senses the nutrient supply to co-ordinate the release of hormones that activate neuronal networks in the brain, leading to the subsequent modulation of hepatic glucose output via the gut-brain-liver axis. The gut hormones also act on multiple peripheral tissues to regulate glucose level through an insulin-dependent and/or -independent mechanism. The brain, especially the hypothalamus, could also response to the hormones such as insulin and leptin and different nutrients to modulate the glucose homeostasis. In this chapter, we review the gut-brain-liver axis and the role of this organ interaction in the control of glucose homeostasis. A better understanding of these pathways will provide novel strategies for improved glycaemic control.

Keywords: glucose homeostasis, hepatic glucose production, gut-brain-liver axis, gastrointestinal hormones, GLP-1, ghrelin, hypothalamus, insulin, leptin, nutrient sensing, lipid sensing

1. Introduction

In human body, glucose production is finely matched to glucose utilisation in order to maintain the glucose level within a relatively narrow range, ~0.8–1.2 mg/dL. This glucose homeostasis is regulated by a network of signals including endocrine, neural and metabolic factors. The various signals monitor the internal energy status and exogenous energy supply to control endogenous glucose production or utilisation.

For a long time, pancreatic insulin and its sensitivity have been the research focus for glucose regulation under normal or diseased condition. Emerging evidence has suggested that the gut and central nervous system (CNS) are also critical in the control of glucose homeostasis. In this chapter, we will review the recent advances on the gut-brain interaction and its effect on the hepatic glucose production (HGP). We will first introduce the concept of 'gut-brain-liver axis' and its physiological relevance in HGP. Then, we will review the current understanding of gastrointestinal (GI) hormones in the control of glucose homeostasis. Lastly, the role of hypothalamic insulin/leptin signalling and nutrients sensing implicated in glucose homeostasis will be discussed.

2. Gut-brain-liver axis

Upon meal ingestion, food enters the digestive tract and triggers a series of mechanical and chemical responses. The tension and stretch receptors, named mechanoreceptors, on the vagal afferent nerves convey these mechanical signals to the brain. On the other hand, the pre-absorbed nutrients and the digested nutrients such as glucose, amino acids and lipids could activate the chemo-receptors on the vagal afferents to regulate glucose homeostasis [1, 2]. Lipids enter the upper intestine in the form of triglycerides (TGs) that are hydrolysed by lipases to form long-chain fatty acids (LCFAs). Intraluminal LCFAs can be further transformed to long chain fatty acyl-co-enzyme A (LCFA-CoA) catalysed by acyl-CoA synthase. Accumulated LCFA-CoAs induce local release of the satiety peptide cholecystokinin (CCK) from the duodenum I cells via a G-protein-coupled receptor GPR40 [3]. Aromatic amino acids could also stimulate the release of CCK via an extracellular calcium-sensing receptor (CaSR) by increasing an intracellular Ca^{2+} level [4, 5]. There are two CCK receptors, CCK-A in the periphery tissues and CCK-B in the brain. Secreted CCK binds to CCK-A receptors on the intestinal vagal afferent neurons and increases the neurophysiological activity and conveys neuronal signals to the CNS [6, 7]. These biochemical signals between the gastrointestinal tract and the CNS are critical for the glucose and lipid metabolism in liver. This communication network between the gut, brain and liver is defined as the gut-brain-liver axis.

2.1. The gut-DVC-liver axis

The main structures in the CNS that are involved in the control of glucose homeostasis are the dorsal vagal complex (DVC) in brain stem and hypothalamus. DVC is composed of the nucleus of the solitary tract (NTS), dorsal motor nucleus of the vagus (DMV) and area postrema. NTS, adjacent to DMV, is the site where the vagal afferent nerves terminate. Activation of vagal afferent signalling increases the NTS neuronal expression of c-Fos, a marker for neuronal activation. Activation of mechanoreceptors in the gut and administration of CCK increases the c-Fos expression in the NTS, suggesting an activation in the gut-DVC signalling [8]. Furthermore, N-methyl D-aspartate (NMDA) channels are located in the NTS neurons. Blocking NTS neuron activity by an NMDA channel blocker abrogates the gut lipid's effect on HGP. This observation further confirms that the NTS neurons relay the signals from the gut to the CNS [2].

NTS neurons are projected to other brain stem nuclei: the nucleus ambiguus (NA) and the DMV. These two nuclei further send out the vagal efferent fibres to form a motor limb of vago-vagal reflexes innervating multiple peripheral organs. Major organs involved in energy consumption and metabolism include the gut, pancreas and liver. Two distinct populations of DMV pre-ganglionic neurons provide parallel innervations to the gut. The cholinergic-muscarinic (excitatory) pathway enhances GI motility and stimulates gastric acid secretion. The non-adrenergic and non-cholinergic (NANC) (inhibitory) pathway inversely decreases the gastric motility. Activation of the vago-vagal reflexes alters the GI functions in different manner via co-operation of these two parallel pathways [9]. In addition, the vagus influences glucose homeostasis through the innervation of the pancreas and liver. Vagal efferents innervating the pancreas enhance the post-prandial insulin release and compensatory insulin responses during prolonged stimulation of pancreatic β cells. If glucose is administered intragastrically that bypasses the vagal activation, post-prandial glucose levels are higher and insulin levels are blunted [10]. Vagal efferents innervating the liver modulate glucose production through the inhibition of key gluconeogenic enzymes and activation of enzymes promoting glycogen synthesis [2, 11].

2.2. The gut-hypothalamus-liver axis

Gastrointestinal signals could be transmitted to the higher centre hypothalamus either through the relay in NTS or via its direct sensing of the gut hormones. Within the hypothalamus, there are three important nuclei, which send out the neural autonomic information to the peripheral organs including the liver. They are ventromedial nucleus (VMN), the lateral hypothalamic area (LHA) and paraventricular nucleus (PVN).

The hypothalamus regulates the hepatic metabolic functions via specific autonomic nerves. Autonomic innervation of liver contains sympathetic and parasympathetic nerves. The pre-ganglionic neurons of sympathetic nerve are located in intermediolateral column of the spinal cord (T7–T12). They convey signals to the post-ganglionic neurons at the celiac and superior mesenteric ganglia and send out the sympathetic splanchnic nerves innervating the liver. The parasympathetic nerves originate from pre-ganglionic neurons in the DMV, while the post-ganglionic neurons have not been clearly identified yet. Neurons in the VMN stimulate liver glycogenolysis (increase liver glucose output) through the activation of the sympathetic system that increases the activity of the liver gluconeogenic enzyme phosphoenolpyruvate carboxykinase (PEPCK) while suppresses pyruvate kinase (PK) activity. The LHA stimulates liver glucogenogenesis (decrease liver glucose output) by decreasing PEPCK activity via the parasympathetic vagal nerve. PVN is a special area that receives projections from both VMN and LHA or other areas. PVN neurons control hepatic glycogen storage via modulating the sympathetic-parasympathetic balance [11, 12].

2.3. Molecular mechanism

The molecular mechanism by which hypothalamic neurons control HGP remains largely unknown. Previous studies have indicated that an intact K_{ATP} channel in the hypothalamic neurons is required for the proper functioning of this brain control on liver glucose metabolism. K_{ATP} channel, a member of the inwardly-rectifying K^+ channel family, is located on the plasma membrane of hypothalamic neurons. It is composed of sulfonylurea receptors (SUR) and inwardly rectifying K^+ channel subunits. Both upper intestinal lipid and central lipid infusion could activate K_{ATP} channel and hyperpolarise the hypothalamic neurons. Activation of K_{ATP} channel in the hypothalamus lowers blood glucose through the inhibition of hepatic gluconeogenesis but not glycogenolysis. The hepatic expressions of *G6Pase* and *Pepck* are significantly reduced. Genetic or pharmacological blockade of hypothalamic K_{ATP} channel, as well as surgical resection of the hepatic vagal nerve, abrogates the gut lipid induced in the inhibition of the HGP [2, 13, 14].

Protein kinase C isoforms also participate in this physiological process. PKCθ is expressed in the hypothalamic neurons, specifically in the neuropeptide Y/agouti-related protein neurons in the arcuate nucleus and the dorsal medial nucleus. Palmitic acid induces the trans-localisation of PKCθ to cell membranes and is associated with impaired insulin and leptin signalling. Knocking down PKCθ in hypothalamic neurons prevented diet-induced obesity and improved insulin signalling [15]. Moreover, activation of PKC in hypothalamus suppresses hepatic glucose production. Inhibition of PKCδ signalling in hypothalamus will abolish this effect, indicating that PKCδ also mediates the central role in glucose homeostasis [16]. In addition to the hypothalamus, PKCδ is expressed in the mucosal layer of duodenum. Activation of duodenal PKCδ stimulates CCK secretion and the gut-brain-liver neuronal pathway and lowers the glucose production in the liver [6, 17].

3. Gut hormones in the regulation of glucose homeostasis

Gastrointestinal tract is the largest endocrine organ in the human body. It consists of a large number of different enteroendocrine cells, which secrete distinct gut hormones. In general, food ingestion acutely stimulates the secretion of gut hormones. These hormones travel in the circulation to the distant tissues, or act locally on vagal afferents and convey signals to CNS. They bind to specific receptors and exert pleotropic metabolic functions through changes in food intake, gastrointestinal motility and energy expenditure.

3.1. GLP-1

Glucagon-like peptide-1 (GLP-1) is a peptide hormone secreted mainly from enteroendocrine L cells of the intestine, and to a much lesser extent from pancreatic α cells and the brain. GLP-1 is cleaved from the precursor proglucagon molecule by protease convertase 1/3. Circulating GLP-1 has two forms, $GLP_{17\text{-}37}$ and $GLP_{17\text{–}36}$ amide. About 80% of circulating GLP1 in human is $GLP_{17\text{–}36}$ amide [18]. Plasma levels of GLP-1 in the basal level range between 5–10 pM and its level could increase to 50 pM after a meal. The half-life of native bioactive GLP-1 is less than 2 min. After secretion from L cells, GLP-1 is quickly excreted by kidney or enzymatically degraded by dipeptidyl peptidase-4 (DPP4) [19]. GLP-1 exerts its biological function through a specific G-protein coupled receptor called GLP-1 receptor (GLP-1r). Using the GLP-1-Cre mice crossed with fluorescent reporter strains, GLP-1 receptor (GLP-1r) is found to be expressed on a variety of tissues including pancreatic β and δ cells, enteric neurons, vagal and dorsal root ganglia, vascular smooth muscle, cardiac atrium and gastric antrum/pylorus [20].

The biological function of endogenous GLP-1 has been examined using GLP-1r antagonists, immunoneutralising antisera and Glp-1r deficient mice. Exendin$_{9-39}$ (Ex-9) is an antagonist of GLP-1r and is the truncated form of the lizard GLP-1-related peptide exendin-4 (Ex-4). Eliminating GLP-1 activity by antisera or infusion of its antagonist results in impaired glucose tolerance and glucose-induced insulin secretion in rodents and humans [21]. Consistently, genetic deficiency of GLP-1r gene in mice causes glucose intolerance and impaired glucose-stimulated insulin secretion, as well as fasting hyperglycemia [22]. Pancreas, especially the pancreatic islet cells, is the target of GLP-1. *In vitro* studies on pancreatic islets or β cell lines demonstrate that GLP-1 increases insulin biosynthesis through both cAMP/PKA-dependent and independent pathways. GLP-1 also stimulates insulin secretion via modulation of K_{ATP} channels, which subsequently increases the intracellular Ca^{2+} level. In addition to the insulinotropic effects, GLP-1 could also stimulate the proliferation of β cell, prevent β cell apoptosis and enhance the differentiation of exocrine-like cells towards a more differentiated β cell phenotype [21]. The signal transduction pathways in which GLP-1 mediates these physiological functions have been extensively studied. Its mitotic effect involves PI3K/AKT, EGFR transactivation, p38 MAPK and PKCζ signalling pathways [23]. Pancreatic and duodenal homeobox gene 1 (PDX1), forkhead box protein O1 (FoxO1), AKT and insulin receptor substrate 2 (IRS2) are essential mediators for the anti-apoptosis effect of GLP-1 [21]. GLP-1 receptor agonist could increase the pancreas weight. However, this effect is not caused by the increase in proliferation of islet or ductal cells. Increased protein synthesis in the exocrine acinar cells accounts for the increase of pancreas weight [24]. GLP-1 also inhibits glucagon secretion from pancreatic α cells after meals to lower the blood glucose. The mechanism by which GLP-1 inhibits glucagon secretion is not completely clear. Recent evidence suggests this effect may be indirectly mediated by another islet hormone, somatostatin [25].

Apart from the pancreas, GLP-1 also acts on CNS to modulate systemic glucose homeostasis. In the brain, only a few neurons in an NTS and area postrema express and secrete GLP-1 [26]. GLP-1 receptors are found in the area postrema, ARC, PVN and VMN [20]. Central administration of GLP-1r antagonist induces glucose intolerance. Under the hyperglycemia condition, activation of central GLP-1 signalling enhances insulin secretion and increases hepatic glycogen storage [27]. Two chronic CNS GLP-1 loss of function mouse models, knocking down GLP-1 gene pre-proglucagon in NTS and chronic ICV infusion of GLP-1r antagonist both caused hyperphagia and glucose intolerance [28]. Therefore, these data suggest that central GLP-1 action plays an important role in systemic glycaemic control.

Liver is another action site of GLP-1. After being secreted from enteroendocrine cells, GLP-1 enters the mesenteric venous and the hepatic portal circulation. Due to the rapid degradation of GLP-1, the close anatomical distance makes the liver as an important target organ for GLP-1. Glucose sensor in the hepatoportal area can sense the circulating glucose level and regulate the hepatic glucose disposal. Basal fasting levels of GLP-1 are sufficient to activate the receptor and maintain its glucoregulatory ability [29]. GLP-1 receptor is present in nodose ganglia and nerve terminals innervating the portal vein, suggesting a possible link of GLP-1 action through the hepatic neuronal network with other peripheral organs. Intraportal infusion of GLP-1 can increase the firing frequency of the vagal afferents and promote glucose disposal in non-hepatic organs [30]. Consistently, administration of the GLP-1r antagonist to block hepatic GLP-1 signalling results in glucose intolerance [31]. *In vitro* study in cultured rat hepatocyte has showed that GLP-1 promotes glycogen accumulation by increasing the activity of glycogen synthase-A while decreasing the activity of glycogen phosphorylase-A [32]. Whether the GLP-1 acts directly on hepatocytes or indirectly through neuronal networks remains to be determined.

Taken together, GLP-1 acts on a variety of distant organs, including pancreas, CNS and liver tissue, to co-ordinate the overall glucose homeostasis.

3.2. Ghrelin

Ghrelin is a 28-amino acids peptide hormone, which is predominantly produced by the enteroendocrine X/A-like cells in the gastric oxyntic mucosa. Ghrelin exerts strong growth hormone (GH)-releasing effects, as well as lactotroph and corticotroph function. It also influences a diverse of biological processes, including food intake, sleep and behaviour, gastrointestinal tract motility and secretion, as well as glucose and energy homeostasis. In this section, we will focus on its role in glucose homeostasis.

Ghrelin is octanoylated at its serine residue catalysed by ghrelin O-acyl transferase (GOAT). This acylation is essential for ghrelin binding to its receptor, GH secretagogue (GHS) receptor 1a (GHS-R1a), to exert its biological function. The GHS-R1a is abundantly expressed in the hypothalamus-pituitary region, in other brain areas and peripheral endocrine and non-endocrine tissues. Binding of GHS-R1a initiates a series of intracellular events, including activation of the phospholipase C signalling, leading to an increased inositol phosphate turnover, inhibition of K^+ channel and increased concentration of intracellular Ca^{2+}.

The first evidence illustrating the relationship of ghrelin to glucose homeostasis is that injection of ghrelin induces a prompt increase of blood glucose, which precedes a transient decrease of insulin secretion in humans [33]. This result suggests that ghrelin-induced blood glucose rise is caused by increased hepatic glucose production and/or peripheral glucose disposal instead of suppressed insulin secretion. Infusion of ghrelin in healthy young men increases plasma glucose and free fatty acid levels. Exogenous ghrelin suppresses glucose-stimulated insulin secretion but demonstrates no effect on basal levels of serum insulin. These observations suggest that ghrelin impairs insulin sensitivity [34, 35]. Evidence from rodent models also consistently demonstrates that exogenous ghrelin administration induces hyperglycaemia and worsens glycaemic control under a high-fat diet condition [36, 37]. Another evidence for the link of ghrelin and glucose homeostasis comes from the clinical assessment of the diurnal variations of ghrelin, glucose and insulin. Fasting ghrelin levels correlates with insulin levels [38]. The ghrelin levels surge before meals and decline after meals. Post-prandial glucose levels are correlated with acyl-ghrelin, independently of insulin. And the ghrelin is lower during sleep than during wake [39].

Since both ghrelin and GHS-R1a are expressed in pancreas, ghrelin synthesised by the islets or elsewhere may regulate the glucose-induced insulin released in autocrine and paracrine manners. Most studies have reported that ghrelin reduces glucose-induced insulin release, while a few studies have found that under certain conditions acyl-ghrelin could amplify insulin secretion [40]. This is partly due to an attenuation of Ca^{2+} signalling [41]. Over-expressing ghrelin in the stomach and brain induces glucose intolerance and attenuates glucose stimulated insulin secretion [42]. Transgenic mice with over-expression of pre-proghrelin gene driven by the rat insulin II promoter-driven (RIP-G Tg) in cells display impaired glucose stimulated insulin secretion. In contrast, over-expression of pre-proghrelin gene driven by rat glucagon promoter-driven (RGP-G Tg) in α cells shows no significant changes in the insulin secretion and glucose metabolism [43]. Ghrelin null mice have no alterations in basal plasma insulin levels but enhanced glucose induced insulin release [44, 45]. High-fat diet-induced glucose intolerance is prevented in ghrelin knockout mice [45]. Isolated ghrelin-deficient islets have increased glucose-induced insulin release but demonstrated no changes in islet density, size and insulin content, suggesting that ghrelin augments insulin secretion through enhanced insulin vesicle exocytosis capacity/efficiency [45]. Detection of ghrelin in the leptin-deficient ob/ob mice also shows enhanced glucose-induced insulin secretion and increased peripheral insulin sensitivity [46].

Despite that ghrelin receptor is not detected, acylated ghrelin increases glucose output in the primary porcine hepatocytes while unacylated ghrelin completely reverses this effect. This result suggests that an acyl-ghrelin probably acts on the primary hepatocytes through a different receptor other than the GHS-R1a and unacylated ghrelin can counteract the acyl-ghrelin's biological function under certain conditions [47]. Ghrelin administration induces a lipogenic and glucogenic pattern of gene expression in the liver and increases hepatic triglyceride content [48]. Also, ghrelin attenuates liver AKT signalling, which may contribute to the blood glucose increments [49].

In CNS, GHS-R1a expression is largely confined to ARC of hypothalamus and somatotroph pituitary cells, both of which are responsible for appetite control and GH release, respectively. Ghrelin is also locally produced by the hypothalamic neurons. These neurons project to NPY/AgRP and POMC neurons, representing a regulatory circuit controlling energy intake [50]. In addition, GH is an important counter-regulatory factor of insulin [51]. Ghrelin may affect glucose homeostasis partly through the GH releasing regulation.

The ghrelin receptors have been detected in the epididymal and pericardial adipose tissue deposits. Ghrelin administration in the mice causes increased adiposity in white adipose tissue and up-regulates UCP1 mRNA expression in brown adipose tissue [52]. *In vitro* study has showed that ghrelin promotes adipogenesis, inhibits lipolysis and glucose uptake into the adipocytes. In addition, ghrelin increases proliferation and differentiation in 3T3-L1 pre-adipocytes. Ghrelin also prevents adipocyte apoptosis caused by serum starvation. Ghrelin activates MAPK and PI3K/AKT pathways in 3T3-L1 pre-adipocytes and adipocytes, whereas inhibition of these pathways blocks the effects of ghrelin on these cells [53, 54].

3.3. Other gut hormones

3.3.1. Peptide YY

Peptide YY (PYY), also known as peptide tyrosine, is a gut peptide hormone primarily released from the enteroendocrine L cells of the colon mucosa. It is also expressed in pancreatic

α cells. Similar to GLP-1, PYY has two forms, the full-length PYY1-36 and a truncated form PYY3-36, which is cleaved by the enzyme DPP IV. The major role of PYY is its inhibition on gut functions, including gastrointestinal tract motility and secretion, gallbladder emptying and food intake [55]. Recent evidence reveals an emerging role of PYY involved in glucose homeostasis. *In vitro* study has showed that PYY inhibits glucose stimulated insulin secretion from the isolated islets by inhibiting the accumulation and action of cAMP [56]. *In vivo* evidence also suggests that deficiency of PYY in mice significantly increases body weight and fat mass. These mice exhibit elevated fasting insulin and glucose stimulated insulin level, suggesting that PYY regulates glucose homeostasis via its stimulation of insulin secretion [57]. Another form of PYY, PYY3-36, elicits no effect on glucose homeostasis under basal or hyperglycaemic conditions. Transgenic mice over-expressing PYY are protected against diet-induced obesity due to increased energy expenditure. Moreover, over-expressing PYY in ob/ob mice significantly reduces adiposity and serum triglyceride levels and improves glucose tolerance [58].

3.3.2. *Nesfatin-1*

Nesfatin-1, an 82-amino acids peptide, is pre-dominantly secreted from gastric X/A-like cells. It was initially identified in the hypothalamus as a satiety molecule [59]. Later, it is also found in pancreatic β cells. Studies using mouse insulinoma cell line MIN6 cells and isolated murine islets have revealed that nesfatin-1 directly enhances the glucose stimulated insulin released by increasing the intracellular Ca^{2+} level through the voltage-dependent L type Ca^{2+} channel [60, 61]. Infusion of nesfatin-1 in the third cerebral ventricle potently inhibits hepatic glucose production and promotes muscle glucose uptake. Inhibition of central nesfatin-1 activity using the adenoviral-mediated RNA interference method increases hepatic glucose flux and decreases glucose uptake in the peripheral tissue. This effect is caused by an alteration of hepatic PEPCK/InsR/IRS-1/AMPK/AKT/mTOR/STAT3 pathway [62, 63]. In addition to the action in CNS, peripheral infusion of nesfatin-1 also significantly improves insulin sensitivity in the skeletal muscle, adipose tissue and liver under normal or high-fat diet condition, via altering AKT phosphorylation and GLUT4 membrane translocation [64].

4. Hypothalamic regulation of glucose homeostasis

Three major hypothalamic nuclei, such as ARC, VMN and PVN, actively participate in the regulation of glucose homeostasis [65]. In 1854, French physiologist Claude Bernard showed that brain lesion in dogs causes hyperglycemia and first proposed that the CNS could be involved in the control of energy and glucose homeostasis. Till date, it has been widely accepted that CNS especially hypothalamus receives and integrates hormonal and nutritional signals from periphery and actively regulates systemic glucose homeostasis. Here, we will focus on how the hypothalamus integrates the critical glucoregulatory hormones, insulin and leptin, as well as the diet nutrients to modulate the intricate regulation of glucose homeostasis.

4.1. Hypothalamic insulin signalling

In addition to the classical insulin target tissues (liver, skeletal muscle and adipose tissue), insulin receptors are also expressed in the CNS, with various densities in different regions [66]. Infusion of insulin in the cerebral ventricle suppresses glucose production independent of circulating insulin and other glucoregulatory hormones [67]. Blunting central insulin signalling by the insulin receptor antisense oligonucleotides also suppresses HGP during hyperinsulinemic clamp studies in rats [68]. Neuron specific knockdown of insulin receptor in whole brain leads to obesity, mild insulin resistance and hyperleptinemia [69]. These results suggest that insulin signalling in hypothalamus controls peripheral glucose production and/or utilisation.

ARC is located at the mediobasal hypothalamus adjacent to the third ventricle and the median eminence, where the blood-brain barrier (BBB) is weakly formed. Insulin or other hormone can easily cross the BBB and functions there. L1 mouse has insulin receptor deficiency in hypothalamus while maintaining insulin receptors in liver and pancreatic β cells. Lack of hypothalamic insulin signalling in L1 mice attenuates the inhibitory effect of insulin on HGP. ARC mainly contains two populations of neurons, the proopiomelanocortin (POMC): expressing neurons and the agouti-related peptide (AgRP) and neuropeptide Y (NPY): co-expressing neurons. To further dissect the role of different groups of neurons, Lin et al. used targeted knock-ins to restore insulin receptor expression in AgRP/NPY or POMC neurons of L1 mice. They have found that re-expression of insulin receptor in AgRP/NPY neurons could normalise insulin's effects on HGP. In contrast, knock-in insulin receptors in POMC neurons increases HGP. These results indicate that these two different types of neurons mediate distinct insulin functions on peripheral glucose and energy homeostasis [70]. Later study has also confirmed that insulin signalling in AgRP neurons is crucial for HGP, expression of G6Pase and insulin-induced expression of interleukin-6 [71].

At the cellular level, the IRS-PI3K pathway mediates the hypothalamic insulin effects on glucose metabolism. Inhibition of PI3K signalling by infusing PI3K inhibitor into the third cerebral ventricle attenuates insulin-induced glucose lowering effects, whereas over-expression of IRS-2 or PI3K substrate AKT enhances the glycaemic response [72].

FoxO1 is a transcriptional repressor of POMC and the transcriptional activator of AgRP. Insulin increases FoxO1 phosphorylation through an InsR/IRS1/PI3K pathway and promotes its exit from the nucleus to the cytoplasm. This translocation dis-inhibits its repression of POMC transcription, allowing for STAT3 binding to the promoter, thus increasing Pomc gene expression. Meanwhile, the nuclear export of FoxO1 decreases *Agrp* gene expression since STAT3 binds to the *Agrp* promoter and inhibits its expression [73]. FoxO1 deletion in AgRP neurons of mice leads to the improved glucose homeostasis and increased sensitivity to insulin and leptin.

K_{ATP} channels are also involved in the hormonal control of glucose metabolism. Activation of K_{ATP} channels in the hypothalamus is sufficient to decrease blood glucose levels by inhibition of HGP. ICV or systemic administration of insulin causes an inhibition of HGP. These effects are prevented by blockade of K_{ATP} channel activity or genetic deletion of SUR1 subunits [14]. Thus, the insulin action in the CNS to regulate glucose level requires intact K_{ATP} channels.

In summary, data from rodents have confirmed that hypothalamic insulin reduces blood glucose concentrations by suppressing gluconeogenesis through inhibition of the expression of key gluconeogenic enzymes. It is worth noting that the role of insulin signalling in glucose homeostasis may vary among species and different physiological conditions. Most experiments have been performed on rodent models. However, it is not exactly the same in other animals. For example, blunting CNS insulin signalling in dogs does not suppress glucose production or gluconeogenesis; instead it augments hepatic glycogen synthesis [74].

4.2. Hypothalamic leptin

Leptin is a hormone produced by an adipose tissue, which regulates pleotropic metabolic functions, such as satiety, body weight gain and glucose homeostasis. Systemic leptin receptor (Lep-r) deficient ob/ob mice are obese, hyperglycemia and hyperinsulinemia. Neuron-specific deletion of Lep-r in mice also exhibits obesity and elevated plasma levels of leptin, glucose and insulin, suggesting the brain is a direct target for leptin and responsible for its role in systemic metabolic effects [75]. Lipodystrophy is a disease characterised by the complete or partial absence of adipose tissue, leptin deficiency, systemic insulin resistance and fatty liver. Central infusion of leptin in the lipodystrophy mice corrects the insulin resistance and improves glucose homeostasis [76]. Overfeeding rapidly induces hepatic, leptin and insulin resistance. Central administration of leptin in overfeeding conscious rats could rescue the hepatic insulin resistance. Leptin inhibits HGP by decreasing glycogenolysis with reduced hepatic expression of G6Pase and PEPCK [77].

Similar to insulin, hypothalamic ARC area is the prominent action site for leptin in regulation of glucose homeostasis. Leptin injection specifically into the ARC increases glucose uptake in BAT [78]. Restoration of Lep-r in the ARC could normalise insulin and blood glucose level without significantly altering the body weight and food intake [79]. Koletsky (fa(k)/fa(k)) rat, in which Lep-r is deficient, is characterised by insulin resistance and hyperglycemia. Re-expression of Lep-r specifically in the ARC region significantly improves peripheral insulin sensitivity and suppresses HGP in this animal. This effect was mediated through CNS-autonomic system -dependent suppression of hepatic expression of the gluconeogenic enzyme, G6Pase and PEPCK [80].

Neuroanatomical and electrophysiological studies have identified POMC neurons as the neuronal target of leptin. Approximately 25–40% of POMC neurons express functional Lep-r. These neurons are distinct from those expressing insulin receptors [81]. Mice lacking leptin signalling in POMC neurons are hyperleptinemic [82]. Simultaneous deletion of insulin and leptin receptors in POMC neurons causes a pronounced insulin resistance [83]. Reconstitution of Lep-r in POMC neurons of the systemic Lep-r deficient mice is sufficient to normalise the hyperglycemia, improve hepatic insulin sensitivity and dyslipidemia [84]. All these results indicate that leptin signalling in POMC neurons plays a key role in regulating glucose homeostasis via control of HGP. The role of leptin in AgRP neurons remains to be investigated.

Lep-r has six splicing variants (Lepr-a to Lepr-f). The long-form Lepr-b is highly expressed and functionally active in the hypothalamus. Leptin binds to the Lepr-b, recruits and activates the

Janus kinase (JAK). JAK in turn phosphorylates Lepr-b and the signal transducer and activator of transcription 3 (STAT3). Phospho-STAT3 traffics into the nucleus and activates the transcription of several genes including suppressor of cytokine signalling 3 (SOCS3). SOCS3 is a negative regulator of leptin signalling. Neuronal deletion of SOCS3 in mice improves leptin sensitivity, with a greater body weight loss and less food intake. These mice are resistant to high-fat diet obesity and hyperleptinemia. The insulin sensitivity is also retained [85]. Lentivirus-mediated knock-down of SOCS3 in hypothalamus significantly increases STAT3 activation, decreases body weight gain and improves metabolic parameters in mice fed with a high-fat diet [86].

Leptin and insulin can engage similar hypothalamic intracellular signalling pathway. Leptin could activate PI3K-PDK1-AKT pathway and influences the neurotransmitter's expression and neuronal activity. Leptin can down-regulate NPY's gene expression and enhance POMC's expression [87, 88]. In addition, leptin causes depolarisation and increased excitability of POMC neurons, while hyperpolarises and inhibits the AgRP neurons [89, 90].

In addition to ARC area, Lep-r is also abundantly expressed in the VMN. Microinjection of leptin in VMN increases glucose uptake in brown adipose tissue, heart, skeletal muscles and spleen through the VMH-sympathetic nervous network [91]. A specific group of neurons in VMN expresses transcription factor steroidogenic factor-1 (SF-1), called SF-1 neurons. Leptin can directly activate SF-1 neurons and leads to body weight loss. Specific deletion of Lep-r in SF-1 neurons in mice results in increased fat mass, hyperinsulinemia and diet-induced obesity, highlighting the importance of VMN in the control of energy homeostasis [92, 93]. SOCS3 in SF1 neurons also negatively regulates leptin signalling, mediating central leptin resistance and glucose dysfunction [93].

4.3. Nutrients sensing and glucose metabolism

4.3.1. Fatty acid/lipid sensing

The brain not only reacts to the circulating hormones but also senses and responds to different nutrients such as lipid, glucose and amino acids. Nutrients sensing in the hypothalamus exert a negative feedback regulation on glucose metabolism and food intake. The most studied example is fatty acid. ICV administration of long-chain fatty acid (LCFA) oleic acid, but not the short-chain fatty acid octanoic acid, could markedly inhibit glucose production and food intake [94]. Fatty acid synthase (FAS) is the rate limiting enzyme for de novo lipogenesis and mediates the synthesis of LCFA. Deletion of FAS in hypothalamus and β cells results in reduced LCFA's level, hypophagic and hypermetabolic in mice fed with a standard or high fat diet [95, 96].

Similar to the gut lipid sensing mechanism, LCFA-CoA is the sensing molecule in the CNS. Malonyl co-enzyme A, a co-enzyme A derivative of malonic acid, plays an important role in chain elongation in LCFA-CoAs biosynthesis. Malonyl-coenzyme A decarboxylase (MCD) mediates the degradation of malonyl co-enzyme A. Thus, over-expression of MCD reduces the abundance of LCFA-CoAs, hence attenuating the lipid sensing ability in the CNS. Indeed, injection of adenovirus expressing MCD into the mediobasal hypothalamus of rats leads to an increase in food intake and weight gain despite increased circulating leptin and insulin [97]. LCFA-CoAs are oxidised in the mitochondria, which is catalysed by two carnitine-dependent long-chain acyl-transferases, known as carnitine palmitoyl transferases (CPT1 and CPT2). CPT1 has two isoforms, liver form of CPT1 (CPT 1A) and muscle form of CPT1 (CPT 1B). Down-regulation of CPT1A expression or pharmacological inhibition of CPT1 activity within the ARC causes an increase in the concentration of LCFA-CoAs and a decrease in the liver glucose output [98]. These central lipid-sensing effects also require intact neuronal K_{ATP} channels and increased hepatic vagal outflow [13], suggesting a mechanism dependent on the brain-liver communication.

4.3.2. Glucose sensing

Our brain is able to directly sense the glucose level [99]. A subset of neurons in hypothalamic nuclei is sensitive to the changes of extracellular glucose level. Some neurons are excited by increased glucose level with more firing frequency while others are inhibited with less firing by same stimuli. The glucose sensitive neurons are located in the VMN, ARC, PVN, as well as the dorsal vagal complex in the brain stem [65]. Central glucose sensing also impacts systemic glucose metabolism. Glucose is converted to lactate, enters the pyruvate metabolism and activates K_{ATP} channels in hypothalamus, leading to an HGP reduction [100]. K_{ATP} channels are able to sense the changes of glucose level and couple membrane excitability to cellular metabolism. High glucose concentrations activate K_{ATP} channel. The intracellular K^+ fluxes out and hyperpolarises the plasma membrane, leading to the subsequent suppression of neurotransmitter release and inhibition of neuronal excitability [101, 102]. Up to date, the identity of neurons that can directly regulate peripheral glucose production is still not known.

4.3.3. Amino acids sensing

Leucine is an essential branched-chain amino acid (BCAA). Recently, it has been demonstrated that leucine gains access to the CNS. It is metabolised to acetyl- and malonyl-CoA in the hypothalamus, which subsequently controls HGP [103]. Another amino acid proline also exerts similar effects on HGP. Proline is a gluconeogenic amino acid, which is metabolised to pyruvate in astrocytes. Pyruvate is eventually metabolised to acetyl- and malonyl-CoA, which in turn regulates HGP. Indeed, increase in either central or systemic proline level acutely lowers blood glucose by decreasing hepatic glycogenolysis as well as gluconeogenesis [104].

4.3.4. Astrocytes in hypothalamus

In addition to neurons, other cell types in the brain, such as astrocytes, contribute to the intricate regulation of central and systemic glucose homeostasis. Circulating glucose is uptake into the CNS via glucose transporter 1 (GLUT-1). GLUT-1 is highly expressed in astrocytes and endothelial cells along the BBB. Recently, astrocytes have been proved to participate in the regulation of hypothalamic glucose sensing and systemic glucose homeostasis. The metabolism and activity of astrocytes are regulated by local hormones and metabolic status. Leptin or nutrients-deprived condition causes glutamate and GLUT-1 changes in the astrocytes, affecting synaptic metabolism, glucose uptake and ultimately neuronal activities [105]. Disruption of insulin signalling in astrocytes affects glucose uptake crossing the blood brain barrier (BBB). It also reduces glucose-induced activation of POMC neurons and impairs hypothalamic mediated suppression of systemic blood glucose [106].

5. Conclusion

Glucose is the basic energy provider for the organisms. Mammalians have redundant systems to control glucose production. Insulin-independent mechanisms account for approximately 50% of overall glucose disposal. Recently, Schwartz et al. have elegantly proposed a two-compartment model which highlights the importance of both pancreatic islet and the brain in glucose homeostasis and development of diabetes [107]. Emerging evidence suggests that the gastrointestinal tract is also a major player in this intricate regulatory process. The best supporting evidence is the pronounced effects of bariatric surgery on obesity and diabetes patients. Although the exact mechanism for the post-operational rapid normalisation of glucose homeostasis is still not clear, it is highly possible that alteration in the gut neuroendocrine restores both insulin-dependent and insulin-independent mechanisms [108]. Understanding the gut-brain-liver axis and its control on glucose homeostasis will provide an alternative strategy for the intervention of diabetes.

Acknowledgements

This work was supported by the National Natural Science Foundation of China (81500619), the Natural Science Foundation of Guangdong Province (2016A030310040), the Shenzhen Science and Technology Project (JCYJ20160422091658982, JCYJ20150324140036854) Shenzhen Peacock Plan (KQTD20140630100746562,827-000107) and the Natural Science Foundation of SZU (201567).

Author details

Zhuo Mao and Weizhen Zhang*

*Address all correspondence to: weizhenz@med.umich.edu

Department of Physiology, Center for Diabetes, Obesity and Metabolism, Shenzhen University Health Science Center, Shenzhen, China

References

[1] Greenberg, D., G.P. Smith, and J. Gibbs, Intraduodenal infusions of fats elicit satiety in sham-feeding rats. Am J Physiol, 1990. **259**(1 Pt 2): p. R110-8.

[2] Wang, P.Y.T., L. Caspi, C.K.L. Lam, M. Chari, X.S. Li, P.E. Light, R. Gutierrez-Juarez, M. Ang, G.J. Schwartz, and T.K.T. Lam, Upper intestinal lipids trigger a gut-brain-liver axis to regulate glucose production. Nature, 2008. **452**(7190): p. 1012-1016.

[3] Liou, A.P., X. Lu, Y. Sei, X. Zhao, S. Pechhold, R.J. Carrero, H.E. Raybould, and S. Wank, The G-protein-coupled receptor GPR40 directly mediates long-chain fatty acid-induced secretion of cholecystokinin. Gastroenterology, 2011. **140**(3): p. 903-12.

[4] Wang, Y., R. Chandra, L.A. Samsa, B. Gooch, B.E. Fee, J.M. Cook, S.R. Vigna, A.O. Grant, and R.A. Liddle, Amino acids stimulate cholecystokinin release through the Ca^{2+}-sensing receptor. Am J Physiol Gastrointest Liver Physiol, 2011. **300**(4): p. G528-37.

[5] Liou, A.P., Y. Sei, X. Zhao, J. Feng, X. Lu, C. Thomas, S. Pechhold, H.E. Raybould, and S.A. Wank, The extracellular calcium-sensing receptor is required for cholecystokinin secretion in response to L-phenylalanine in acutely isolated intestinal I cells. Am J Physiol Gastrointest Liver Physiol, 2011. **300**(4): p. G538-46.

[6] Breen, D.M., J.T. Yue, B.A. Rasmussen, A. Kokorovic, G.W. Cheung, and T.K. Lam, Duodenal PKC-delta and cholecystokinin signaling axis regulates glucose production. Diabetes, 2011. **60**(12): p. 3148-53.

[7] Cheung, G.W.C., A. Kokorovic, C.K.L. Lam, M. Chari, and T.K.T. Lam, Intestinal cholecystokinin controls glucose production through a neuronal network. Cell Metab, 2009. **10**(2): p. 99-109.

[8] Rinaman, L., G.E. Hoffman, E.M. Stricker, and J.G. Verbalis, Exogenous cholecystokinin activates cFos expression in medullary but not hypothalamic neurons in neonatal rats. Brain Res Dev Brain Res, 1994. **77**(1): p. 140-5.

[9] Travagli, R.A., G.E. Hermann, K.N. Browning, and R.C. Rogers, Musings on the wanderer: what's new in our understanding of vago-vagal reflexes? III. Activity-dependent plasticity in vago-vagal reflexes controlling the stomach. Am J Physiol Gastrointest Liver Physiol, 2003. **284**(2): p. G180-7.

[10] Teff, K.L., Visceral nerves: vagal and sympathetic innervation. J Parenteral Enteral Nutr, 2008. **32**(5): p. 569-571.

[11] Shimazu, T., Regulation of glycogen metabolism in liver by the autonomic nervous system. V. Activation of glycogen synthetase by vagal stimulation. Biochem Biophys Acta, 1971. **252**(1): p. 28-38.

[12] Uyama, N., A. Geerts, and H. Reynaert, Neural connections between the hypothalamus and the liver. Anat Record A Discov Mol Cell Evol Biol, 2004. **280A**(1): p. 808-820.

[13] Lam, T.K., A. Pocai, R. Gutierrez-Juarez, S. Obici, J. Bryan, L. Aguilar-Bryan, G.J. Schwartz, and L. Rossetti, Hypothalamic sensing of circulating fatty acids is required for glucose homeostasis. Nat Med, 2005. **11**(3): p. 320-7.

[14] Pocai, A., T.K. Lam, R. Gutierrez-Juarez, S. Obici, G.J. Schwartz, J. Bryan, L. Aguilar-Bryan, and L. Rossetti, Hypothalamic K(ATP) channels control hepatic glucose production. Nature, 2005. **434**(7036): p. 1026-31.

[15] Benoit, S.C., C.J. Kemp, C.F. Elias, W. Abplanalp, J.P. Herman, S. Migrenne, A.L. Lefevre, C. Cruciani-Guglielmacci, C. Magnan, F. Yu, K. Niswender, B.G. Irani, W.L. Holland, and D.J. Clegg, Palmitic acid mediates hypothalamic insulin resistance by altering PKC-theta subcellular localization in rodents. J Clin Invest, 2009. **119**(9): p. 2577-89.

[16] Ross, R., P.Y.T. Wang, M. Chari, C.K.L. Lam, L. Caspi, H. Ono, E.D. Muse, X. Li, R. Gutierrez-Juarez, P.E. Light, G.J. Schwartz, L. Rossetti, and T.K.T. Lam, Hypothalamic protein kinase C regulates glucose production. Diabetes, 2008. **57**(8): p. 2061-2065.

[17] Kokorovic, A., G.W. Cheung, D.M. Breen, M. Chari, C.K. Lam, and T.K. Lam, Duodenal mucosal protein kinase C-delta regulates glucose production in rats. Gastroenterology, 2011. **141**(5): p. 1720-7.

[18] Orskov, C., L. Rabenhoj, A. Wettergren, H. Kofod, and J.J. Holst, Tissue and plasma concentrations of amidated and glycine-extended glucagon-like peptide I in humans. Diabetes, 1994. **43**(4): p. 535-9.

[19] Hansen, L., C.F. Deacon, C. Orskov, and J.J. Holst, Glucagon-like peptide-1-(7-36) amide is transformed to glucagon-like peptide-1-(9-36)amide by dipeptidyl peptidase IV in the capillaries supplying the L cells of the porcine intestine. Endocrinology, 1999. **140**(11): p. 5356-63.

[20] Richards, P., H.E. Parker, A.E. Adriaenssens, J.M. Hodgson, S.C. Cork, S. Trapp, F.M. Gribble, and F. Reimann, Identification and characterization of GLP-1 receptor-expressing cells using a new transgenic mouse model. Diabetes, 2014. **63**(4): p. 1224-1233.

[21] Drucker, D.J., The role of gut hormones in glucose homeostasis. J Clin Invest, 2007. **117**(1): p. 24-32.

[22] Scrocchi, L.A., T.J. Brown, N. MaClusky, P.L. Brubaker, A.B. Auerbach, A.L. Joyner, and D.J. Drucker, Glucose intolerance but normal satiety in mice with a null mutation in the glucagon-like peptide 1 receptor gene. Nat Med, 1996. **2**(11): p. 1254-8.

[23] Buteau, J., S. Foisy, C.J. Rhodes, L. Carpenter, T.J. Biden, and M. Prentki, Protein kinase czeta activation mediates glucagon-like peptide-1-induced pancreatic beta-cell proliferation. Diabetes, 2001. **50**(10): p. 2237-43.

[24] Koehler, J.A., L.L. Baggio, X. Cao, T. Abdulla, J.E. Campbell, T. Secher, J. Jelsing, B. Larsen, and D.J. Drucker, Glucagon-like peptide-1 receptor agonists increase pancreatic mass by induction of protein synthesis. Diabetes, 2015. **64**(3): p. 1046-56.

[25] de Heer, J., C. Rasmussen, D.H. Coy, and J.J. Holst, Glucagon-like peptide-1, but not glucose-dependent insulinotropic peptide, inhibits glucagon secretion via somatostatin (receptor subtype 2) in the perfused rat pancreas. Diabetologia, 2008. **51**(12): p. 2263-70.

[26] Yamamoto, H., T. Kishi, C.E. Lee, B.J. Choi, H. Fang, A.N. Hollenberg, D.J. Drucker, and J.K. Elmquist, Glucagon-like peptide-1-responsive catecholamine neurons in the area postrema link peripheral glucagon-like peptide-1 with central autonomic control sites. J Neurosci, 2003. **23**(7): p. 2939-46.

[27] Knauf, C., P.D. Cani, C. Perrin, M.A. Iglesias, J.F. Maury, E. Bernard, F. Benhamed, T. Gremeaux, D.J. Drucker, C.R. Kahn, J. Girard, J.F. Tanti, N.M. Delzenne, C. Postic, and R. Burcelin, Brain glucagon-like peptide-1 increases insulin secretion and muscle insulin resistance to favor hepatic glycogen storage. J Clin Invest, 2005. **115**(12): p. 3554-63.

[28] Barrera, J.G., K.R. Jones, J.P. Herman, D.A. D'Alessio, S.C. Woods, and R.J. Seeley, Hyperphagia and increased fat accumulation in two models of chronic CNS glucagon-like peptide-1 loss of function. J Neurosci, 2011. **31**(10): p. 3904-13.

[29] Burcelin, R., A. Da Costa, D. Drucker, and B. Thorens, Glucose competence of the hepatoportal vein sensor requires the presence of an activated glucagon-like peptide-1 receptor. Diabetes, 2001. **50**(8): p. 1720-8.

[30] Nakabayashi, H., M. Nishizawa, A. Nakagawa, R. Takeda, and A. Niijima, Vagal hepatopancreatic reflex effect evoked by intraportal appearance of tGLP-1. Am J Physiol, 1996. **271**(5 Pt 1): p. E808-13.

[31] Vahl, T.P., M. Tauchi, T.S. Durler, E.E. Elfers, T.M. Fernandes, R.D. Bitner, K.S. Ellis, S.C. Woods, R.J. Seeley, J.P. Herman, and D.A. D'Alessio, Glucagon-like peptide-1 (GLP-1) receptors expressed on nerve terminals in the portal vein mediate the effects of endogenous GLP-1 on glucose tolerance in rats. Endocrinology, 2007. **148**(10): p. 4965-4973.

[32] Valverde, I., M. Morales, F. Clemente, M.I. Lopez-Delgado, E. Delgado, A. Perea, and M.L. Villanueva-Penacarrillo, Glucagon-like peptide 1: a potent glycogenic hormone. FEBS Lett, 1994. **349**(2): p. 313-6.

[33] Broglio, F., E. Arvat, A. Benso, C. Gottero, G. Muccioli, M. Papotti, A.J. van der Lely, R. Deghenghi, and E. Ghigo, Ghrelin, a natural GH secretagogue produced by the stomach, induces hyperglycemia and reduces insulin secretion in humans. J Clin Endocrinol Metab, 2001. **86**(10): p. 5083-6.

[34] Vestergaard, E.T., T.K. Hansen, L.C. Gormsen, P. Jakobsen, N. Moller, J.S. Christiansen, and J.O. Jorgensen, Constant intravenous ghrelin infusion in healthy young men: clinical pharmacokinetics and metabolic effects. Am J Physiol Endocrinol Metab, 2007. **292**(6): p. E1829-36.

[35] Tong, J., R.L. Prigeon, H.W. Davis, M. Bidlingmaier, S.E. Kahn, D.E. Cummings, M.H. Tschop, and D. D'Alessio, Ghrelin suppresses glucose-stimulated insulin secretion and deteriorates glucose tolerance in healthy humans. Diabetes, 2010. **59**(9): p. 2145-51.

[36] Nieminen, P. and A.M. Mustonen, Effects of peripheral ghrelin on the carbohydrate and lipid metabolism of the tundra vole (Microtus oeconomus). Gen Comp Endocrinol, 2004. **138**(2): p. 182-7.

[37] Asakawa, A., A. Inui, T. Kaga, G. Katsuura, M. Fujimiya, M.A. Fujino, and M. Kasuga, Antagonism of ghrelin receptor reduces food intake and body weight gain in mice. Gut, 2003. **52**(7): p. 947-52.

[38] Purnell, J.Q., D.S. Weigle, P. Breen, and D.E. Cummings, Ghrelin levels correlate with insulin levels, insulin resistance, and high-density lipoprotein cholesterol, but not with gender, menopausal status, or cortisol levels in humans. J Clin Endocrinol Metab, 2003. **88**(12): p. 5747-52.

[39] Spiegel, K., E. Tasali, R. Leproult, N. Scherberg, and E. Van Cauter, Twenty-four-hour profiles of acylated and total ghrelin: relationship with glucose levels and impact of time of day and sleep. J Clin Endocrinol Metab, 2011. **96**(2): p. 486-93.

[40] Delhanty, P.J., and A.J. van der Lely, Ghrelin and glucose homeostasis. Peptides, 2011. **32**(11): p. 2309-18.

[41] Dezaki, K., H. Hosoda, M. Kakei, S. Hashiguchi, M. Watanabe, K. Kangawa, and T. Yada, Endogenous ghrelin in pancreatic islets restricts insulin release by attenuating Ca^{2+} signaling in beta-cells: implication in the glycemic control in rodents. Diabetes, 2004. **53**(12): p. 3142-51.

[42] Bewick, G.A., A. Kent, D. Campbell, M. Patterson, M.A. Ghatei, S.R. Bloom, and J.V. Gardiner, Mice with hyperghrelinemia are hyperphagic and glucose intolerant and have reduced leptin sensitivity. Diabetes, 2009. **58**(4): p. 840-6.

[43] Iwakura, H., K. Hosoda, C. Son, J. Fujikura, T. Tomita, M. Noguchi, H. Ariyasu, K. Takaya, H. Masuzaki, Y. Ogawa, T. Hayashi, G. Inoue, T. Akamizu, H. Hosoda, M. Kojima, H. Itoh, S. Toyokuni, K. Kangawa, and K. Nakao, Analysis of rat insulin II promoter-ghrelin transgenic mice and rat glucagon promoter-ghrelin transgenic mice. J Biol Chem, 2005. **280**(15): p. 15247-56.

[44] Morand, C., C. Remesy, and C. Demigne, Regulation of glycogen metabolism in liver. Diabetes Metab, 1992. **18**(1 Pt 2): p. 87-95.

[45] Dezaki, K., H. Sone, M. Koizumi, M. Nakata, M. Kakei, H. Nagai, H. Hosoda, K. Kangawa, and T. Yada, Blockade of pancreatic islet-derived ghrelin enhances insulin secretion to prevent high-fat diet-induced glucose intolerance. Diabetes, 2006. **55**(12): p. 3486-93.

[46] Sun, Y., M. Asnicar, P.K. Saha, L. Chan, and R.G. Smith, Ablation of ghrelin improves the diabetic but not obese phenotype of ob/ob mice. Cell Metab, 2006. **3**(5): p. 379-86.

[47] Gauna, C., P.J. Delhanty, L.J. Hofland, J.A. Janssen, F. Broglio, R.J. Ross, E. Ghigo, and A.J. van der Lely, Ghrelin stimulates, whereas des-octanoyl ghrelin inhibits, glucose output by primary hepatocytes. J Clin Endocrinol Metab, 2005. **90**(2): p. 1055-60.

[48] Barazzoni, R., A. Bosutti, M. Stebel, M.R. Cattin, E. Roder, L. Visintin, L. Cattin, G. Biolo, M. Zanetti, and G. Guarnieri, Ghrelin regulates mitochondrial-lipid metabolism gene expression and tissue fat distribution in liver and skeletal muscle. Am J Physiol Endocrinol Metab, 2005. **288**(1): p. E228-35.

[49] Barazzoni, R., M. Zanetti, M.R. Cattin, L. Visintin, P. Vinci, L. Cattin, M. Stebel, and G. Guarnieri, Ghrelin enhances in vivo skeletal muscle but not liver AKT signaling in rats. Obesity (Silver Spring), 2007. **15**(11): p. 2614-23.

[50] Cowley, M.A., R.G. Smith, S. Diano, M. Tschop, N. Pronchuk, K.L. Grove, C.J. Strasburger, M. Bidlingmaier, M. Esterman, M.L. Heiman, L.M. Garcia-Segura, E.A. Nillni, P. Mendez, M.J. Low, P. Sotonyi, J.M. Friedman, H. Liu, S. Pinto, W.F. Colmers, R.D. Cone, and T.L. Horvath, The distribution and mechanism of action of ghrelin in the CNS demonstrates a novel hypothalamic circuit regulating energy homeostasis. Neuron, 2003. **37**(4): p. 649-61.

[51] Vijayakumar, A., R. Novosyadlyy, Y. Wu, S. Yakar, and D. LeRoith, Biological effects of growth hormone on carbohydrate and lipid metabolism. Growth Horm IGF Res, 2010. **20**(1): p. 1.

[52] Tsubone, T., T. Masaki, I. Katsuragi, K. Tanaka, T. Kakuma, and H. Yoshimatsu, Ghrelin regulates adiposity in white adipose tissue and UCP1 mRNA expression in brown adipose tissue in mice. Regul Pept, 2005. **130**(1-2): p. 97-103.

[53] Kim, M.S., C.Y. Yoon, P.G. Jang, Y.J. Park, C.S. Shin, H.S. Park, J.W. Ryu, Y.K. Pak, J.Y. Park, K.U. Lee, S.Y. Kim, H.K. Lee, Y.B. Kim, and K.S. Park, The mitogenic and antiapoptotic actions of ghrelin in 3T3-L1 adipocytes. Mol Endocrinol, 2004. **18**(9): p. 2291-301.

[54] Miegueu, P., D. St Pierre, F. Broglio, and K. Cianflone, Effect of desacyl ghrelin, obestatin and related peptides on triglyceride storage, metabolism and GHSR signaling in 3T3-L1 adipocytes. J Cell Biochem, 2011. **112**(2): p. 704-14.

[55] Boey, D., A. Sainsbury, and H. Herzog, The role of peptide YY in regulating glucose homeostasis. Peptides, 2007. **28**(2): p. 390-395.

[56] Nieuwenhuizen, A.G., S. Karlsson, T. Fridolf, and B. Ahren, Mechanisms underlying the insulinostatic effect of peptide yy in mouse pancreatic-islets. Diabetologia, 1994. **37**(9): p. 871-878.

[57] Boey, D., S. Lin, T. Karl, P. Baldock, N. Lee, R. Enriquez, M. Couzens, K. Slack, R. Dallmann, A. Sainsbury, and H. Herzog, Peptide YY ablation in mice leads to the development of hyperinsulinaemia and obesity. Diabetologia, 2006. **49**(6): p. 1360-70.

[58] Boey, D., S. Lin, R.F. Enriquez, N.J. Lee, K. Slack, M. Couzens, P.A. Baldock, H. Herzog, and A. Sainsbury, PYY transgenic mice are protected against diet-induced and genetic obesity. Neuropeptides, 2008. **42**(1): p. 19-30.

[59] Oh, I.S., H. Shimizu, T. Satoh, S. Okada, S. Adachi, K. Inoue, H. Eguchi, M. Yamamoto, T. Imaki, K. Hashimoto, T. Tsuchiya, T. Monden, K. Horiguchi, M. Yamada, and M. Mori,

Identification of nesfatin-1 as a satiety molecule in the hypothalamus. Nature, 2006. **443**(7112): p. 709-12.

[60] Nakata, M., K. Manaka, S. Yamamoto, M. Mori, and T. Yada, Nesfatin-1 enhances glucose-induced insulin secretion by promoting Ca(2+) influx through L-type channels in mouse islet beta-cells. Endocr J, 2011. **58**(4): p. 305-13.

[61] Gonzalez, R., B.K. Reingold, X. Gao, M.P. Gaidhu, R.G. Tsushima, and S. Unniappan, Nesfatin-1 exerts a direct, glucose-dependent insulinotropic action on mouse islet beta- and MIN6 cells. J Endocrinol, 2011. **208**(3): p. R9–R16.

[62] Yang, M., Z. Zhang, C. Wang, K. Li, S. Li, G. Boden, L. Li, and G. Yang, Nesfatin-1 action in the brain increases insulin sensitivity through Akt/AMPK/TORC2 pathway in diet-induced insulin resistance. Diabetes, 2012. **61**(8): p. 1959-68.

[63] Wu, D., M. Yang, Y. Chen, Y. Jia, Z.A. Ma, G. Boden, L. Li, and G. Yang, Hypothalamic nesfatin-1/NUCB2 knockdown augments hepatic gluconeogenesis that is correlated with inhibition of mTOR-STAT3 signaling pathway in rats. Diabetes, 2014. **63**(4): p. 1234-47.

[64] Li, Z., L. Gao, H. Tang, Y. Yin, X. Xiang, Y. Li, J. Zhao, M. Mulholland, and W. Zhang, Peripheral effects of nesfatin-1 on glucose homeostasis. PLoS One, 2013. **8**(8): p. e71513.

[65] Roh, E., D.K. Song, and M.-S. Kim, Emerging role of the brain in the homeostatic regulation of energy and glucose metabolism. Exp Mol Med, 2016. **48**: p. e216.

[66] Havrankova, J., J. Roth, and M. Brownstein, Insulin receptors are widely distributed in the central nervous system of the rat. Nature, 1978. **272**(5656): p. 827-829.

[67] Obici, S., B.B. Zhang, G. Karkanias, and L. Rossetti, Hypothalamic insulin signaling is required for inhibition of glucose production. Nat Med, 2002. **8**(12): p. 1376-82.

[68] Obici, S., Z. Feng, G. Karkanias, D.G. Baskin, and L. Rossetti, Decreasing hypothalamic insulin receptors causes hyperphagia and insulin resistance in rats. Nat Neurosci, 2002. **5**(6): p. 566-72.

[69] Bruning, J.C., D. Gautam, D.J. Burks, J. Gillette, M. Schubert, P.C. Orban, R. Klein, W. Krone, D. Muller-Wieland, and C.R. Kahn, Role of brain insulin receptor in control of body weight and reproduction. Science, 2000. **289**(5487): p. 2122-25.

[70] Lin, H.V., L. Plum, H. Ono, R. Gutierrez-Juarez, M. Shanabrough, E. Borok, T.L. Horvath, L. Rossetti, and D. Accili, Divergent regulation of energy expenditure and hepatic glucose production by insulin receptor in agouti-related protein and POMC neurons. Diabetes, 2010. **59**(2): p. 337-346.

[71] Konner, A.C., R. Janoschek, L. Plum, S.D. Jordan, E. Rother, X.S. Ma, C. Xu, P. Enriori, B. Hampel, G.S. Barsh, C.R. Kahn, M.A. Cowley, F.M. Ashcroft, and J.C. Bruning, Insulin action in AgRP-expressing neurons is required for suppression of hepatic glucose production. Cell Metab, 2007. **5**(6): p. 438-449.

[72] Gelling, R.W., G.J. Morton, C.D. Morrison, K.D. Niswender, M.G. Myers Jr, C.J. Rhodes, and M.W. Schwartz, Insulin action in the brain contributes to glucose lowering during insulin treatment of diabetes. Cell Metab, 2006. **3**(1): p. 67-73.

[73] Kim, M.S., Y.K. Pak, P.G. Jang, C. Namkoong, Y.S. Choi, J.C. Won, K.S. Kim, S.W. Kim, H.S. Kim, J.Y. Park, Y.B. Kim, and K.U. Lee, Role of hypothalamic Foxo1 in the regulation of food intake and energy homeostasis. Nat Neurosci, 2006. **9**(7): p. 901-906.

[74] Ramnanan, C.J., V. Saraswathi, M.S. Smith, E.P. Donahue, B. Farmer, T.D. Farmer, D. Neal, P.E. Williams, M. Lautz, A. Mari, A.D. Cherrington, and D.S. Edgerton, Brain insulin action augments hepatic glycogen synthesis without suppressing glucose production or gluconeogenesis in dogs. J Clin Invest, 2011. **121**(9): p. 3713-3723.

[75] Cohen, P., C. Zhao, X. Cai, J.M. Montez, S.C. Rohani, P. Feinstein, P. Mombaerts, and J.M. Friedman, Selective deletion of leptin receptor in neurons leads to obesity. J Clin Invest, 2001. **108**(8): p. 1113-21.

[76] Asilmaz, E., P. Cohen, M. Miyazaki, P. Dobrzyn, K. Ueki, G. Fayzikhodjaeva, A.A. Soukas, C.R. Kahn, J.M. Ntambi, N.D. Socci, and J.M. Friedman, Site and mechanism of leptin action in a rodent form of congenital lipodystrophy. J Clin Invest, 2004. **113**(3): p. 414-24.

[77] Pocai, A., K. Morgan, C. Buettner, R. Gutierrez-Juarez, S. Obici, and L. Rossetti, Central leptin acutely reverses diet-induced hepatic insulin resistance. Diabetes, 2005. **54**(11): p. 3182-9.

[78] Toda, C., T. Shiuchi, S. Lee, M. Yamato-Esaki, Y. Fujino, A. Suzuki, S. Okamoto, and Y. Minokoshi, Distinct effects of leptin and a melanocortin receptor agonist injected into medial hypothalamic nuclei on glucose uptake in peripheral tissues. Diabetes, 2009. **58**(12): p. 2757-65.

[79] Coppari, R., M. Ichinose, C.E. Lee, A.E. Pullen, C.D. Kenny, R.A. McGovern, V. Tang, S.M. Liu, T. Ludwig, S.C. Chua, B.B. Lowell, and J.K. Elmquist, The hypothalamic arcuate nucleus: a key site for mediating leptin's effects on glucose homeostasis and locomotor activity. Cell Metab, 2005. **1**(1): p. 63-72.

[80] German, J., F. Kim, G.J. Schwartz, P.J. Havel, C.J. Rhodes, M.W. Schwartz, and G.J. Morton, Hypothalamic Leptin signaling regulates hepatic insulin sensitivity via a neurocircuit involving the vagus nerve. Endocrinology, 2009. **150**(10): p. 4502-4511.

[81] Balthasar, N., R. Coppari, J. McMinn, S.M. Liu, C.E. Lee, V. Tang, C.D. Kenny, R.A. McGovern, S.C. Chua, Jr., J.K. Elmquist, and B.B. Lowell, Leptin receptor signaling in POMC neurons is required for normal body weight homeostasis. Neuron, 2004. **42**(6): p. 983-91.

[82] Hill, J.W., C.F. Elias, M. Fukuda, K.W. Williams, E.D. Berglund, W.L. Holland, Y.R. Cho, J.C. Chuang, Y. Xu, M. Choi, D. Lauzon, C.E. Lee, R. Coppari, J.A. Richardson, J.M. Zigman, S. Chua, P.E. Scherer, B.B. Lowell, J.C. Bruning, and J.K. Elmquist, Direct

insulin and leptin action on pro-opiomelanocortin neurons is required for normal glucose homeostasis and fertility. Cell Metab, 2010. **11**(4): p. 286-97.

[83] Huo, L., K. Gamber, S. Greeley, J. Silva, N. Huntoon, X.H. Leng, and C. Bjorbaek, Leptin-dependent control of glucose balance and locomotor activity by POMC neurons. Cell Metab, 2009. **9**(6): p. 537-47.

[84] Berglund, E.D., C.R. Vianna, J. Donato Jr, M.H. Kim, J.C. Chuang, C.E. Lee, D.A. Lauzon, P. Lin, L.J. Brule, M.M. Scott, R. Coppari, and J.K. Elmquist, Direct leptin action on POMC neurons regulates glucose homeostasis and hepatic insulin sensitivity in mice. J Clin Invest, 2012. **122**(3): p. 1000-9.

[85] Mori, H., R. Hanada, T. Hanada, D. Aki, R. Mashima, H. Nishinakamura, T. Torisu, K.R. Chien, H. Yasukawa, and A. Yoshimura, Socs3 deficiency in the brain elevates leptin sensitivity and confers resistance to diet-induced obesity. Nat Med, 2004. **10**(7): p. 739-43.

[86] Liu, Z.J., J. Bian, Y.L. Zhao, X. Zhang, N. Zou, and D. Li, Lentiviral vector-mediated knockdown of SOCS3 in the hypothalamus protects against the development of diet-induced obesity in rats. Diabetes Obes Metab, 2011. **13**(10): p. 885-92.

[87] Schwartz, M.W., D.G. Baskin, T.R. Bukowski, J.L. Kuijper, D. Foster, G. Lasser, D.E. Prunkard, D. Porte Jr, S.C. Woods, R.J. Seeley, and D.S. Weigle, Specificity of leptin action on elevated blood glucose levels and hypothalamic neuropeptide Y gene expression in ob/ob mice. Diabetes, 1996. **45**(4): p. 531-5.

[88] Schwartz, M.W., R.J. Seeley, S.C. Woods, D.S. Weigle, L.A. Campfield, P. Burn, and D.G. Baskin, Leptin increases hypothalamic pro-opiomelanocortin mRNA expression in the rostral arcuate nucleus. Diabetes, 1997. **46**(12): p. 2119-23.

[89] Hill, J.W., K.W. Williams, C. Ye, J. Luo, N. Balthasar, R. Coppari, M.A. Cowley, L.C. Cantley, B.B. Lowell, and J.K. Elmquist, Acute effects of leptin require PI3K signaling in hypothalamic proopiomelanocortin neurons in mice. J Clin Invest, 2008. **118**(5): p. 1796-805.

[90] van den Top, M., K. Lee, A.D. Whyment, A.M. Blanks, and D. Spanswick, Orexigen-sensitive NPY/AgRP pacemaker neurons in the hypothalamic arcuate nucleus. Nat Neurosci, 2004. **7**(5): p. 493-4.

[91] Minokoshi, Y., M.S. Haque, and T. Shimazu, Microinjection of leptin into the ventromedial hypothalamus increases glucose uptake in peripheral tissues in rats. Diabetes, 1999. **48**(2): p. 287-91.

[92] Dhillon, H., J.M. Zigman, C. Ye, C.E. Lee, R.A. McGovern, V. Tang, C.D. Kenny, L.M. Christiansen, R.D. White, E.A. Edelstein, R. Coppari, N. Balthasar, M.A. Cowley, S. Chua Jr, J.K. Elmquist, and B.B. Lowell, Leptin directly activates SF1 neurons in the VMH, and this action by leptin is required for normal body-weight homeostasis. Neuron, 2006. **49**(2): p. 191-203.

[93] Bingham, N.C., K.K. Anderson, A.L. Reuter, N.R. Stallings, and K.L. Parker, Selective loss of leptin receptors in the ventromedial hypothalamic nucleus results in increased adiposity and a metabolic syndrome. Endocrinology, 2008. **149**(5): p. 2138-48.

[94] Obici, S., Z. Feng, K. Morgan, D. Stein, G. Karkanias, and L. Rossetti, Central administration of oleic acid inhibits glucose production and food intake. Diabetes, 2002. **51**(2): p. 271-5.

[95] Chakravarthy, M.V., Y. Zhu, M. Lopez, L. Yin, D.F. Wozniak, T. Coleman, Z. Hu, M. Wolfgang, A. Vidal-Puig, M.D. Lane, and C.F. Semenkovich, Brain fatty acid synthase activates PPAR alpha to maintain energy homeostasis. J Clin Invest, 2007. **117**(9): p. 2539-2552.

[96] Chakravarthy, M.V., Y. Zhu, L. Yin, T. Coleman, K.L. Pappan, C.A. Marshall, M.L. McDaniel, and C.F. Semenkovich, Inactivation of hypothalamic FAS protects mice from diet-induced obesity and inflammation. J. Lipid Res, 2009. **50**(4): p. 630-640.

[97] He, W., T.K.T. Lam, S. Obici, and L. Rossetti, Molecular disruption of hypothalamic nutrient sensing induces obesity. Nat Neurosci, 2006. **9**(2): p. 227-233.

[98] Pocai, A., T.K. Lam, S. Obici, R. Gutierrez-Juarez, E.D. Muse, A. Arduini, and L. Rossetti, Restoration of hypothalamic lipid sensing normalizes energy and glucose homeostasis in overfed rats. J Clin Invest, 2006. **116**(4): p. 1081-91.

[99] Oomura, Y., T. Ono, H. Ooyama, and M.J. Wayner, Glucose and Osmosensitive Neurones of the Rat Hypothalamus. Nature, 1969. **222**(5190): p. 282-284.

[100] Lam, T.K., R. Gutierrez-Juarez, A. Pocai, and L. Rossetti, Regulation of blood glucose by hypothalamic pyruvate metabolism. Science, 2005. **309**(5736): p. 943-7.

[101] Plum, L., B.F. Belgardt, and J.C. Bruning, Central insulin action in energy and glucose homeostasis. J Clin Invest, 2006. **116**(7): p. 1761-6.

[102] Amoroso, S., H. Schmidantomarchi, M. Fosset, and M. Lazdunski, Glucose, sulfonylureas, and neurotransmitter release: role of ATP-sensitive K+ channels. Science, 1990. **247**(4944): p. 852-854.

[103] Su, Y., T.K.T. Lam, W. He, A. Pocai, J. Bryan, L. Aguilar-Bryan, and R. Gutierrez-Juarez, Hypothalamic leucine metabolism regulates liver glucose production. Diabetes, 2012. **61**(1): p. 85-93.

[104] Arrieta-Cruz, I., Y. Su, C.M. Knight, T.K.T. Lam, and R. Gutiérrez-Juárez, Evidence for a role of proline and hypothalamic astrocytes in the regulation of glucose metabolism in rats. Diabetes, 2013. **62**(4): p. 1152-1158.

[105] Fuente-Martin, E., C. Garcia-Caceres, M. Granado, M.L. de Ceballos, M.A. Sanchez-Garrido, B. Sarman, Z.W. Liu, M.O. Dietrich, M. Tena-Sempere, P. Argente-Arizon, F. Diaz, J. Argente, T.L. Horvath, and J.A. Chowen, Leptin regulates glutamate and glucose transporters in hypothalamic astrocytes. J Clin Invest, 2012. **122**(11): p. 3900-3913.

[106] García-Cáceres, C., C. Quarta, L. Varela, Y. Gao, T. Gruber, B. Legutko, M. Jastroch, P. Johansson, J. Ninkovic, C.X. Yi, O. Le Thuc, K. Szigeti-Buck, W. Cai, C. W. Meyer, P. T. Pfluger, A. M. Fernandez, S. Luquet, Stephen C. Woods, I. Torres-Alemán, C.R. Kahn, M. Götz, Tamas L. Horvath, and Matthias H. Tschöp, Astrocytic Insulin Signaling Couples Brain Glucose Uptake with Nutrient Availability. Cell, 2016. **166**(4): p. 867-880.

[107] Schwartz, M.W., R.J. Seeley, M.H. Tschop, S.C. Woods, G.J. Morton, M.G. Myers, and D. D'Alessio, Cooperation between brain and islet in glucose homeostasis and diabetes. Nature, 2013. **503**(7474): p. 59-66.

[108] Cote, C.D., M. Zadeh-Tahmasebi, B.A. Rasmussen, F.A. Duca, and T.K. Lam, Hormonal signaling in the gut. J Biol Chem, 2014. **289**(17): p. 11642-9.

Chapter 5

Blood Glucose Prediction for "Artificial Pancreas" System

Nikolay A. Bazaev and Kirill V. Pozhar

Additional information is available at the end of the chapter

http://dx.doi.org/10.5772/67142

Abstract

The aim of modern science in diabetes therapy is to develop a closed-loop system to control blood glucose (BG) ("artificial pancreas"). Such a system includes glucose monitor, insulin pump and algorithms of their interaction and blood glucose (BG) dynamics analysis. Current work is devoted to mathematic modeling of BG dynamics, development of BG prediction algorithm and its approbation on clinical data Diabetes Research in Children Network (DirecNet). Prediction algorithm is based on sigma model of BG regulation and is supposed to be used for "artificial pancreas" system. The algorithm was tested under condition of possible errors in glucometer and/or insulin pump operation. Root-mean-square error of BG prediction is 15.7 mg/dl. Algorithm corrects 97.5% errors.

Keywords: artificial pancreas, mathematical model, blood glucose dynamics, prediction, noninvasive, glucose meter, glucometer, diabetes mellitus

1. Introduction

Diabetes mellitus is one of the most widespread and serious diseases. Diabetes mellitus type 1 means that the pancreas loses its ability to produce insulin. The aim of modern science in diabetes therapy is to develop a closed-loop system («Artificial Pancreas»), which can compensate such an insufficiency. «Artificial Pancreas» is a system combining a glucose sensor, a control algorithm and an insulin infusion device [1]. In the case of diabetes mellitus (type 2), pump insulin therapy is not relevant due to metabolism disorder, whereas in the case of insulin-dependent one (type 1), it is the most effective method of diabetes compensation.

Modern systems for blood glucose (BG) monitoring concentration and insulin pumps for subcutaneous insulin injection have certain potential for raising the level of diabetes mellitus therapy. However, any errors in such a system can cause dangerous consequences for the patient's health: hypo- or hyperglycemia.

Some errors occur due to biomechanics of the sensor-tissue area (e.g., mechanical pressure in the sensor implantation zone) [2]. In clinical practice, there are two types of errors associated with glucose monitoring: abrupt fluctuations of BG and short-term decrease in the BG sensor sensitivity. Incorrect functioning of BG sensor can be associated with failure in insulin injection rate caused by tubing occlusion.

Moreover insulin and glucose have delays in absorption and action (about 30–100 min for insulin and 5–15 min for glucose) [1]. Thus, insulin should be injected in the way to prevent hyperglycemia.

2. Modeling of blood glucose dynamics

2.1. Mathematic models of blood glucose and insulin dynamics

In the human body, glucose is the main and the most multifunctional energy source for metabolic processes. Due to utmost importance of stable blood glucose level, there is a complicated regulation system. The scheme of blood glucose and insulin regulation is represented in **Figure 1**.

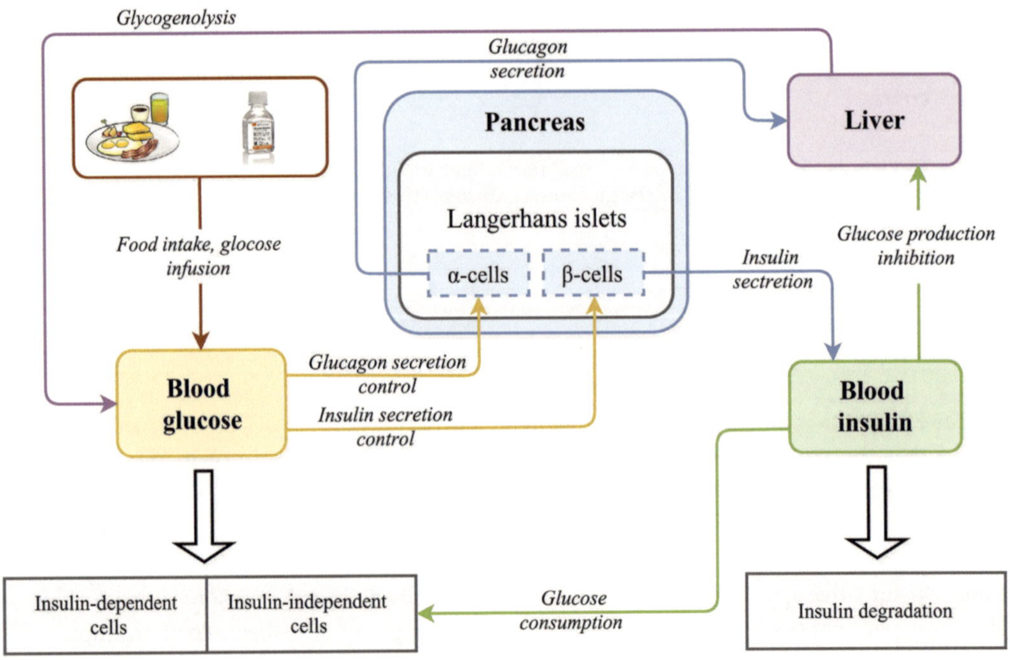

Figure 1. Scheme of glucose and insulin regulation in the human body.

Most BG dynamics models tend to describe physiological processes of glucose and insulin regulation. The simplest model describing BG dynamics is the Bergman's minimal model [3]. It is a system of three ordinary differential equations (ODE) (Eq. (1)):

http://dx.doi.org/10.5772/67142

$$\begin{cases} \dfrac{dg(t)}{dt} = -\Big(b_1 + x(t)\Big)g(t) + b_1 g_b \\ \dfrac{dx(t)}{dt} = -b_2 x(t) + b_3\Big(i(t) - i_b\Big) \\ \dfrac{di(t)}{dt} = b_4\Big(g(t) - b_5\Big)^{+} t - b_6\Big(i(t) - i_b\Big) \end{cases} \tag{1}$$

where $g(t)$ is blood glucose level; $i(t)$ is blood insulin level; $x(t)$ is proportional to insulin in the remote compartment; $g(b)$ is normal BG; $i(b)$ is basal insulin level; b_1, b_2, b_3, b_4 and b_6 are rate constants; and b_5 is glucose threshold level (b_5).

The advantage of the model (Eq. (1)) is in proper description of BG and insulin dynamics. However, under some circumstances, it behaves incorrectly. The model describes BG peak with returning to the initial BG level g_b.

Many models are based on Bergman's minimal model but take into consideration some more parameters [4, 5].

Sturis' model [6] includes two negative feedbacks, describing glucose absorption due to insulin action and insulin secretion as a reaction to BG rising. The model is described by the following ODE system (Eq. (2)):

$$\begin{cases} \dfrac{dg(t)}{dt} = g_{in} - f_2(g(t)) - f_3(g(t))f_4(i_i(t)) + f_5(x_3(t)) \\ \dfrac{di_p(t)}{dt} = f_1(g(t)) - e\left(\dfrac{i_p(t)}{v_p} - \dfrac{i_i(t)}{v_i}\right) - \dfrac{i_p(t)}{t_p} \\ \dfrac{di_i(t)}{dt} = e\left(\dfrac{i_p(t)}{v_p} - \dfrac{i_i(t)}{v_i}\right) - \dfrac{i_p(t)}{t_i} \\ \dfrac{dx_1(t)}{dt} = \dfrac{3}{v_d}(i_p(t) - x_1(t)) \\ \dfrac{dx_2(t)}{dt} = \dfrac{3}{v_d}(x_1(t) - x_2(t)) \\ \dfrac{dx_3(t)}{dt} = \dfrac{3}{v_d}(x_2(t) - x_3(t)), \end{cases} \tag{2}$$

where g_{in} is exogenous glucose infusion rate; $i_p(t)$ is plasma insulin; $i_i(t)$ is interstitial insulin; $f_1(z)$ is insulin secretion (function of glucose); $f_2(z)$ is insulin-independent glucose utilization (function of glucose); $f_3(z)$ is insulin-dependent glucose utilization (function of glucose); $f_4(z)$ is insulin-dependent glucose utilization (function of insulin); $f_5(z)$ is glucose production (function of insulin); x_1, x_2 and x_3 are variables representing delay process between plasma insulin and glucose production; t_1 and t_2 are time constants of plasma and interstitial insulin degradation, respectively; t_3 is delay time between plasma insulin and glucose production; v_p and v_i are volumes of plasma and interstitial insulin, respectively; v_d is volume of glucose space; and e is rate constant for exchange of insulin between the two compartments.

Engelborghs' model [7] is based on Sturis' model (Eq. (1)) and takes into account time delay of insulin secretion and of BG ascent. Insulin injections can be described in the following way (Eq. (3)) (with an assumption that it equals patient's pancreas insulin):

$$\begin{cases} \frac{dg(t)}{dt} = g_{in} - f_2\left(g(t)\right) - f_3\left(g(t)\right)f_4\left(i(t)\right) + f_5\left(i(t-\tau_2)\right) \\ \frac{di(t)}{dt} = \alpha f_1\left(g(t)\right) - \frac{i(t)}{t_1} + (1-\alpha)f_1\left(g(t-\tau_1)\right), \end{cases} \tag{3}$$

where $\alpha f_1\left(g(t)\right)$ is the fraction of insulin delivered by the pancreas of a normal subject into the circulation to maintain blood sugar at its physiological level, τ_1 is pancreatic insulin production time delay and τ_2 is hepatic glucose production delay. In contrast with Sturis' model, Engelborghs' model is sustainable.

Bennett-Gourley's model (D.L. Bennett, S.A. Gourley) [8] for glucose and insulin dynamics uses the following differential equations (Eq. (4)):

$$\begin{cases} \frac{di_p}{dt} = f_1\left(g(t)\right) - e\left(\frac{i_p(t)}{v_p} - \frac{i_i(t)}{v_i}\right) - \frac{i_p(t)}{t_p} \\ \frac{di_i(t)}{dt} = e\left(\frac{i_p(t)}{v_p} - \frac{i_i(t)}{v_i}\right) - \frac{i_i(t)}{t_i} \\ \frac{dg(t)}{dt} = g_{in} - f_2\left(g(t)\right) - qg(t)f_4\left(i_i(t)\right) + f_5\left(i_p(t-\tau)\right), \end{cases} \tag{4}$$

where $q > 0$ is constant. The Bennett-Gourley's model has no sustained oscillations.

Despite significant international experience in modeling of blood glucose and insulin dynamics, there are a number of individual peculiarities of patient that effect BG like his/her physical activity, stresses, etc. Possible ways of taking into account such a variety of factors are in using of indistinct logic and neural networks. As input data for neural networks, information could be used that can have any influence on BG dynamics. The output data are blood glucose concentration or amount of insulin to infuse to normalize BG.

However, processing of neural network algorithms is even more consuming than systems of ordinary differential equations of high orders.

2.2. Sigma model

Direct application of mathematic model for BG dynamics is in its predicting capabilities. On its basis, one can make an algorithm for BG prediction that can be used as a part of "artificial pancreas" system. Mathematic models on the basis of high-order differential equations are rather complicated to be used in a processing unit of a portable glucometer, especially when they are complexificated with additional data about numerous factors effecting blood glucose concentration. That is why simple empirical mathematic models [9] for such calculations are more preferable.

http://dx.doi.org/10.5772/67142

Let us consider independently BG ascent and descent processes. The main factor effecting BG increase is food intake. Food can be characterized by amount of carbohydrates and glycemic index. Thus, food intake can be described by three parameters: m_f is food mass, v is relative carbohydrate coefficient and GI is glycemic index. It should be noted that mentioned parameters have to be estimated for each product independently. Carbohydrate mass influences the amplitude of BG increase, whereas glycemic index defines the rate of glucose appearance in the blood stream.

For BG analysis precise track form during BG ascent is not important, only amplitude and assimilation time. The function of accumulating glucose in the blood can be expressed in the following way (Eq. (5)):

$$G(t) = G_0 + \sum_i \frac{m_{f_i} . v_i}{km_T} \cdot \left(\frac{1}{1 + \exp\left(-\alpha\left(t - t_i - \tau_{gluc} - \frac{\tau_1}{\overline{GI_i}}\right) / \frac{\tau_1}{\overline{GI_i}}\right)} \right), \quad (5)$$

where $G(t)$ is BG value K, G_0 is initial BG value, $\overline{GI_i} = GI_i/100$ is normalized glycemic index, α is normalization coefficient, τ_1 is time of penetration of all glucose into the blood, t_i is time of food intake and τ_{gluc} is time delay between food intake and glucose penetration into the blood.

Function of insulin effect on BG can be described as follows (Eq. (6)):

$$G(t) = G_0 - \sum_j \frac{\mu \cdot n_j}{km_T} \cdot \left(\frac{1}{1 + \exp\left(- \beta(t - t_j - \tau_{ins} - \tau_2)/\tau_2\right)} \right), \quad (6)$$

where β is normalization coefficient, τ_2 is time of insulin action spike, t_j is time of insulin injection, τ_{ins} is delay between insulin injection and its action, n_j is the dose of injected insulin and μ is coefficient that refers to the glucose descent due to one unit of injected insulin.

The model is based on logistic function or sigma function. That is why it is call sigma model.

General equation of BG dynamics that takes into consideration food intake and insulin injection can be described as superposition of functions (Eqs. (5) and (6)):

$$\begin{aligned} G(t) = G_0 + \sum_i \frac{m_{f_i} \cdot v_i}{km_T} \cdot \left(\frac{1}{1 + \exp\left(-\alpha\left(t - t_i - \tau_{gluc} - \frac{\tau_1}{\overline{GI_i}}\right) / \frac{\tau_1}{\overline{GI_i}}\right)} \right) - , \\ - \sum_j \frac{\mu \cdot n_j}{km_T} \cdot \left(\frac{1}{1 + \exp\left(- \beta(t - t_j - \tau_{ins} - \tau_2)/\tau_2\right)} \right) \end{aligned} \quad (7)$$

Eq. (7) is the simplest variant of BG dynamics model. However, there are other independent factors affecting BG dynamics:

–Glucose elimination by kidneys

–Glucose accumulation in the liver in the form of glycogen

–Glycogenolysis driven by glucagon

–Natural residual insulin secretion by pancreas β-cells

–Gluconeogenesis

In current approach, we do not take into account glucose accumulation and glycogenolysis in the human body. These processes do not change BG rapidly and can be evaluated from blood glucose measurements.

As shown in **Figure 2**, the model (Eq. (7)) describes ascent and descent of BG track quite accurately. It should be noted that model provides better results in the diabetes mellitus type 1 case, because it is characterized by very little or even no insulin secretion. Sigma model appears to be applicable for BG prediction.

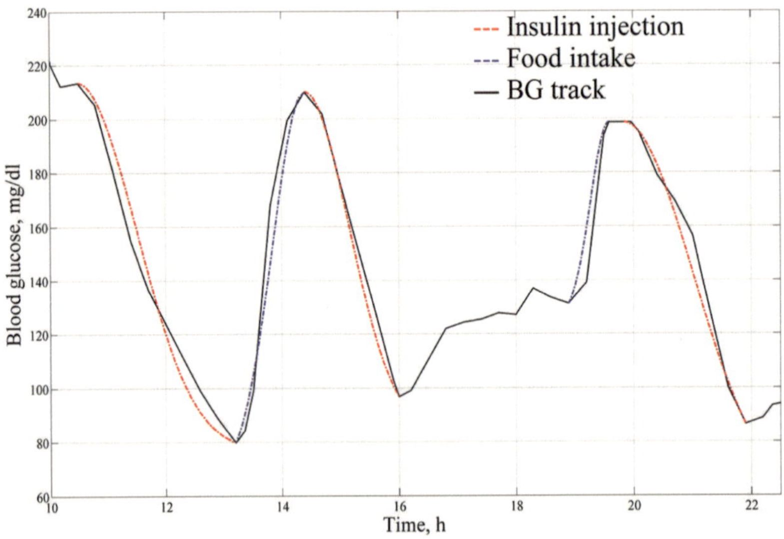

Figure 2. Modeling of blood glucose dynamics with sigma model.

2.3. BG profile database

For experimental evaluation of developed mathematic model and the prediction algorithm, we need BG tracks with information about factors that affect BG dynamics. Protocols of Diabetes Research in Children Network (DirecNet) are used for that purpose. DirecNet investigates the potential use of glucose monitoring technologies and its impact on the compensation of type 1 diabetes in children [10]. Investigations with completed protocols were directed to estimate accuracy of constant glucose monitor system (CGMS) [11–14], especially of FreeStyle Navigator sensor [15–17] and evaluate physical activities on BG [18, 19] and basal insulin injection [20] influence on hypoglycemia in children with type 1 diabetes.

http://dx.doi.org/10.5772/67142

Experimental results are represented as a number of tables, reports about patient anamnesis and received data. Data protocols include detailed description of experiments. Tables contain the following information: BG measurements, insulin injections, initial patient data, calibrating BG tracks, anamnesis, meal, physical activity, dawn phenomenon, blood counts (CBC, hormones, biochemical, clinical BG), medicines, stress, etc. Database includes more than 4000 records about food intake, insulin injections and physical activity and more than 2000 glucose daily tracks. This available data can be used for approbation of mathematic models and prediction algorithm.

DirecNet data was structured, analyzed and filtered; daily BG tracks, provided by different glucose meters and clinical blood analysis, were synchronized (**Figure 3**).

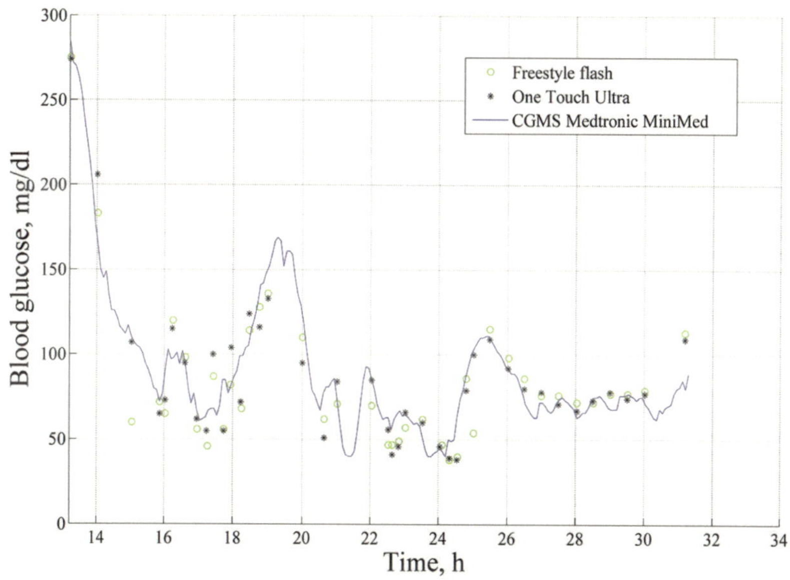

Figure 3. Combined blood glucose track, provided by glucometers OneTouch Ultra, FreeStyleFlash and Medtronic CGMS.

In **Figure 4**, there is a BG track with information about food intake and insulin injection. BG growth and reduction accurately correlate with food and insulin intake time and amount.

Clinical analyses were taken as the most precise. Measurement results provided by noninvasive glucometers FreeStyle Flash and One Touch Ultra (measurements were performed once in a 30–60 minutes) match with each other and in the case of no clinical data, their arithmetic mean value was taken as a true one. Supporting track was formed according to these data. Medtronic CGMS and FreeStyle Navigator data (measurements were performed once in a 1–5 minutes) are less accurate and have noticeable value drift; however, due to higher digitalization, they make it possible to reconstruct precise BG track form. Tracks provided by glucose monitors were corrected in accordance with a supporting track. As a result, a set of BG tracks with higher digitalization (1–5 min step) and accuracy was obtained (**Figure 5**).

Obtained data were used for approbation mathematic models and prediction algorithm.

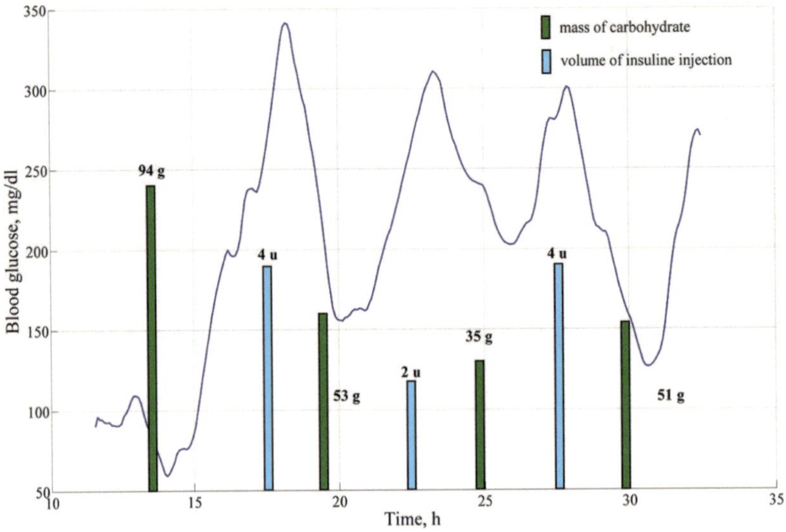

Figure 4. BG track with food intake and insulin injection data.

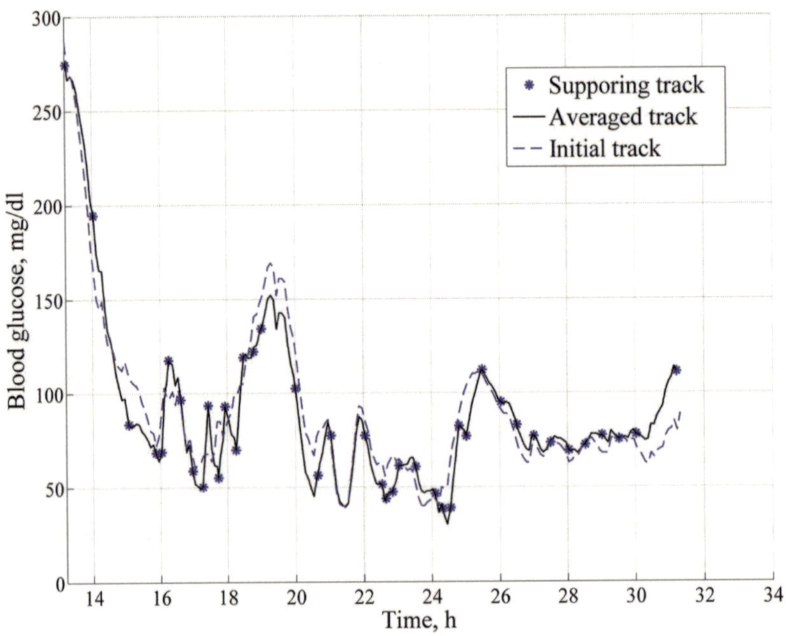

Figure 5. Averaged blood glucose concentration track.

2.4. Mathematic model approbation

DirecNet databases give enough information for sigma model approbation. However, it is necessary to define a wide range of parameters for other models of BG dynamics.

In particular, Sturis', Engelborghs' and Bennett-Gourley's models require such parameters as quantities of plasma insulin, interstitial insulin and glucose, pancreatic insulin production, insulin-independent glucose utilization, insulin-dependent glucose utilization, the influence of insulin on hepatic glucose production, distribution volumes for insulin in plasma and interstitial fluid.

These parameters were obtained from BG tracks using a neural network. Bennett-Gourley's model has broader set of parameters in comparison with Sturis' and Engelborghs' models. The purpose of the neural network was to reconstruct Bennett-Gourley's model parameters by an input track. Neural network was trained in the following way. Randomly generated set of parameters for Bennett-Gourley's model (from physiological range of values) was used to reconstruct BG track. This track was included in training sample as a stimuli vector and model parameters as a response vector. After training the neural network, it was used to define necessary parameters for models by analyzing track from DirecNet. Neural network processing was implemented by Matlab Neural Net Toolbox using FeedForwardNet, Elman Recurrent Neural Network and embedded neural network. The best efficiency was demonstrated by training algorithm "trainbr" with 14 neurons in the hidden layer and activation functions "logsig" and "purelin" combination. Average error of model parameter determination was 14%. Elman Recurrent Neural Net provided the best accuracy; RMSE was equaled to 27 mg/dl.

It should be noted that track representation by all models appeared to be more precise when neural network was trained by Bennett-Gourley's model. The model parameters are presented in **Table 1** and its description is represented in the work of Tolic et al. [21].

We compared results of track modeling by Engelborghs', Bennett-Gourley's and sigma model parameters with BG tracks for T1DM patients from DirecNet databases. Missing parameters of first three models were defined by neural network; sigma model was built according to parameters specified in experiment protocols. Results are represented in **Figure 6**. Moments of food intake and insulin injections with corresponding doses are also marked on the plot.

As it turned out, Sturis' model is highly oscillating and has significant dips on the track after food intake. Engelborghs', Bennett-Gourley's and sigma models describe BG track form rather accurately; nevertheless, due to the fact that these models show better results in the case of healthy subjects, it can be noticed that BG tends to return to initial value. Moreover, sigma model is more accurate in describing BG track. Model error in the physiological blood glucose concentration range was equal to 15.7 mg/dl. It makes the model applicable for BG prediction algorithm development on its bases. It should be noted that presented RMSE is average for the whole day. The accuracy is higher for short-term prediction (30 min) and model iterative correction.

The disadvantage of the model is its ability to predict only growth or reduction of BG induced by food intake or insulin injection, respectively. As there are a lot of other BG-affecting factors [22], sigma model can be extended and improved but is applicable for prediction algorithm development as is.

Symbol	Value	Dimension
V_g	10	l
R_m	210	mmol/min
a_1	300	mg/l
C_1	2000	mg/l
U_b	72	mg/min
C_2	144	mg/l
C_3	1000	mg/l
V_p	3	l
V_i	11	l
E	0.2	l/min
U_0	40	mg/min
U_m	940	mg/min
β	1.77	–
C_4	80	mol/l
R_g	180	mg/min
α	0.29	l/mol
C_5	26	l/mol
t_i	100	min
t_p	6	min
t_d	36	min

Table 1. Model parameters for neural net training.

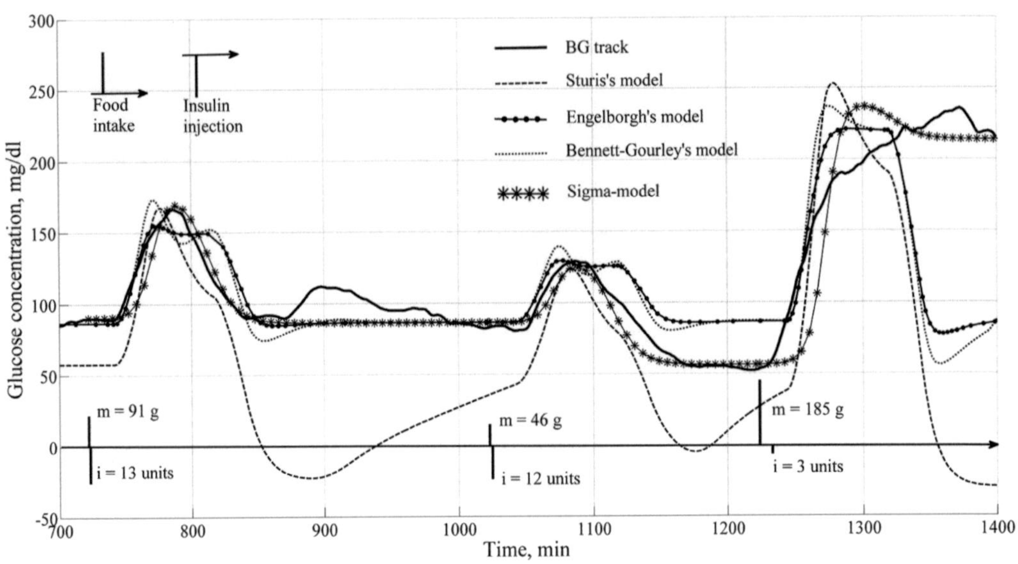

Figure 6. Comparison between BGC tracks based on experimental data and model.

3. Predictive algorithm

BG prediction algorithm is needed for adjusting the rate of insulin injection and detection of insulin pump or glucometer failure.

Algorithm is based on sigma model [23] and calculates BG dynamics for 2 hours in advance. Initial BG level is measured by invasive glucose meter in the fasting state. On receiving new data (from glucose monitor or from patient), a new prediction is made and previous predictions are erased. After each measurement made by BG monitor, the model analyses received data. Incorrect system operation cases and algorithm response on that are described below. If there are no errors in measurement process, new BG dynamic prediction is performed on the base of received data. Linear shift of predicted track toward measured value is calculated as follows (Eq. (8)):

$$\Delta G = |G_{gluc} - G| \cdot \frac{\delta_m}{\delta_m + \delta_{gluc}}, \tag{8}$$

where δ_{gluc} is glucose monitor value error and δ_m is averaged error of mathematic model for the period between two measurements.

In order to evaluate measured BG adequacy, we consider two types of prediction: short-term and long-term predictions, to detect local errors and systematic ones, respectively.

3.1. Error detecting and correction

The first type of errors is related to possible spikes due to BG monitor software or hardware failure during BG measurements. Such kind of errors is revealed to single value of glucometer data that differ significantly from predicted one. Threshold for taking measured value as falling can be expressed as (Eq. (9))

$$\delta_\Sigma = \frac{2 \cdot \delta_m + \delta_{gluc}}{2} + \delta_{tol} \tag{9}$$

where δ_Σ is the total tolerance and δ_{tol} is the theoretical tolerance of BG data deviation from values obtained by the model when accurate measurement and modeling occur.

When such an error takes place, algorithm ignores the data from glucose monitor and considers predicted data as true. Error signalization is not obligatory in this case. During the period of seven measurements, only three falling out values are allowed. If incorrect values appear, too often algorithm makes a report about possible system failure. Algorithm operation example with such type of errors is represented in **Figure 7**.

The second error type is connected to damage (split or disconnection) or cross clamping of insulin pump infusion set. Because of undelivered insulin, BG systematically increases. If long-term prognosis differs from glucometer data with insulin being infused, the algorithm sends a signal about possible error and suggests patient to check an infusion set (**Figure 8**).

It also may happen that a patient for some reason did not enter meal data. In this case, as well as in the previous one, significant deviation of glucometer measurements from long-term prediction occurs. If insulin was not infused, algorithm sends a signal about the error (**Figure 9**) and suggests patient to confirm that he/she didn't have a meal during last 60 minutes.

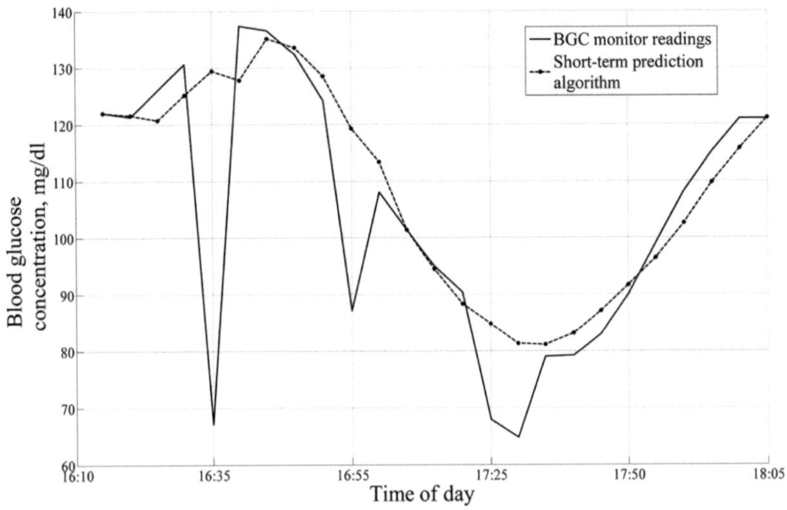

Figure 7. Example of BG track spikes elimination by predictive algorithm.

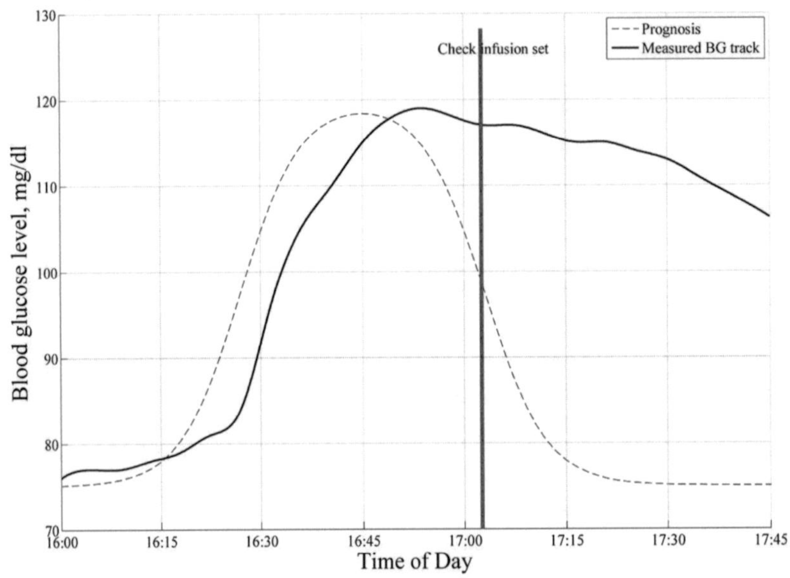

Figure 8. Algorithm operation in the case of infusion system failure.

In the case of sharp reduction of sensor sensibility, insulin pump breakdown, or other system operation failure, systematical deviations occur and there is no possibility to eliminate them by prior mentioned methods. Then algorithm reports about system failure, turns off automatic insulin infusion and suggests a patient to consult his/her doctor.

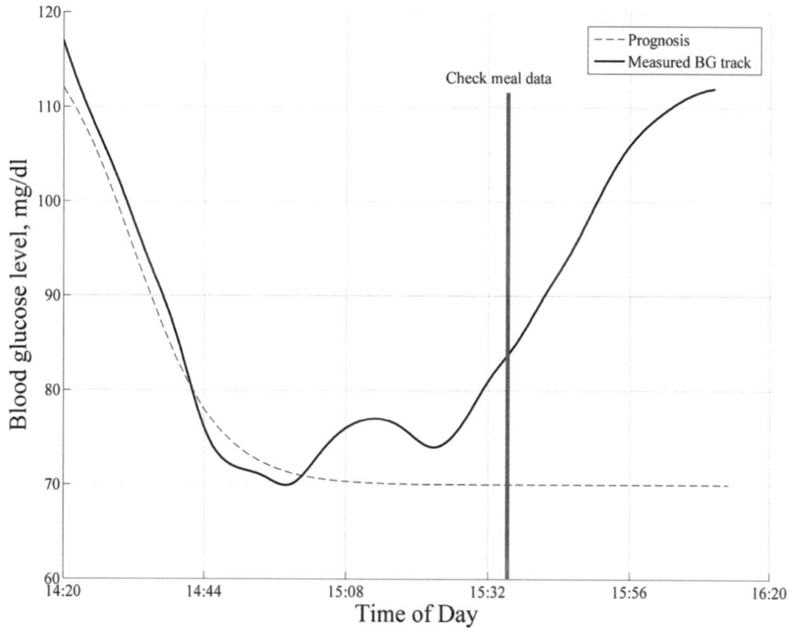

Figure 9. Algorithm operation when patient didn't enter meal data.

3.2. Insulin dosage calculation

The aim of insulin pump is automatic insulin infusion in two regimes: basal and bolus. The pancreas produces insulin in advance so that BG would be in certain range of values. There are several schemes for individual insulin injections and in most cases, it should be injected before BG meal. Rapid-acting insulin should be infused straight after the meal but before blood glucose increase. Moreover, patients often need basal insulin infusion twice a day, but with insulin pump, it is infused continuously [24]. Bolus insulin dose is calculated in accordance to meal composition and product glycemic index.

Using such data and sigma model, one can calculate BG dynamics and evaluate dose and time of insulin injection to make BG normal after food assimilation. As a rule a standard value is used, for instance, 80 mg/dl. As sigma model is sensible only to food intake and insulin injection, BG with $t \to \infty$ tends to a constant (Eq. (10)):

$$\left(\frac{m_{f_i \cdot v_i}}{km_T} \cdot \left(\frac{1}{1+\exp\left(-\alpha\left(t-t_i-\tau_{gluc}-\frac{\tau_1}{GI_i}\right)/\frac{\tau_1}{GI_i}\right)}\right)\right)\Bigg|_{t\to\infty} = \frac{m_{f_i \cdot v_i}}{km_T} \cdot \left(\frac{1}{1+\exp(-\infty)}\right) = \frac{m_{f_i \cdot v_i}}{km_T} \tag{10}$$

In order to reach desired BG value G_{norm} (Eq. (11)), it is necessary to infuse the following insulin volume n_{norm} (Eq. (12)):

$$G_{norm} = \left(G_0 + \sum_i \frac{m_{f_i \cdot v_i}}{km_T} - \sum_j \frac{\mu \cdot n_j}{km_T} \right) - \frac{\mu \cdot n_{norm}}{km_T}, \tag{11}$$

$$n_{norm} = \frac{km_T}{\mu} \left(\left(G_0 + \sum_i \frac{m_{f_i \cdot v_i}}{km_T} - \sum_j \frac{\mu \cdot n_j}{km_T} \right) - G_{norm} \right) \tag{12}$$

Insulin infusion time depends on type of insulin used in the insulin pump. Selection of insulin infusion pattern is performed according to definite kind of insulin. The pattern is selected to minimize BG deviation from normal range of values. It is important to prevent hypoglycemic values of BG, because lack of insulin can be compensated by additional bolus, but extra BG reduction caused by insulin over-infusion cannot be balanced automatically.

Predicting track is shifted depending on measured BG values. As a result, it may happen that real BG value after some time will be lower than it was initially predicted. Meanwhile a new prediction will contain a zone where predicting value will appear in hypoglycemia area. Insulin and food have assimilation delays. If insulin of extra value is infused, it will be impossible to register and correct appropriate BG changes and a patient will have hypoglycemia after a while. In order to avoid hypoglycemia, it was offered to infuse only 75% of bolus value at once and the rest after increase of BG.

3.3. Concept of predictive algorithm work principles

General concept of predictive algorithm operation is represented by the flowchart (**Figure 10**).

3.4. Experimental tests

Algorithm was approbated using BG tracks from DirecNet database.

Algorithm successfully detected cases of meal data absence. In the case further data input and algorithm operation were impossible to perform. As all tests were performed under the medical supervision, DirecNet protocols have no data about possible insulin infusion system failure.

However, measured BG tracks had many spikes. Results of such kind of error elimination are represented in **Table 2**.

Algorithm corrected 97.5% of BG monitor errors. Meanwhile only 3.7% of correct BG monitor data was detected as errors. Generally, measurements were taken as errors in the ranges of BG sharp ascent and descent due to food intake or insulin infusion, respectively. That is explained by high track differential. Having that, little variations of delays in food uptake and in insulin infusion lead to significant difference of predicted and real BG values. Uncorrected errors of BG appeared, when spikes had deviated toward the predicted track and then had fallen into algorithm's correct values zone. Along with real errors of BG monitor, the synthetic errors were included in tracks. Approbation of the

http://dx.doi.org/10.5772/67142

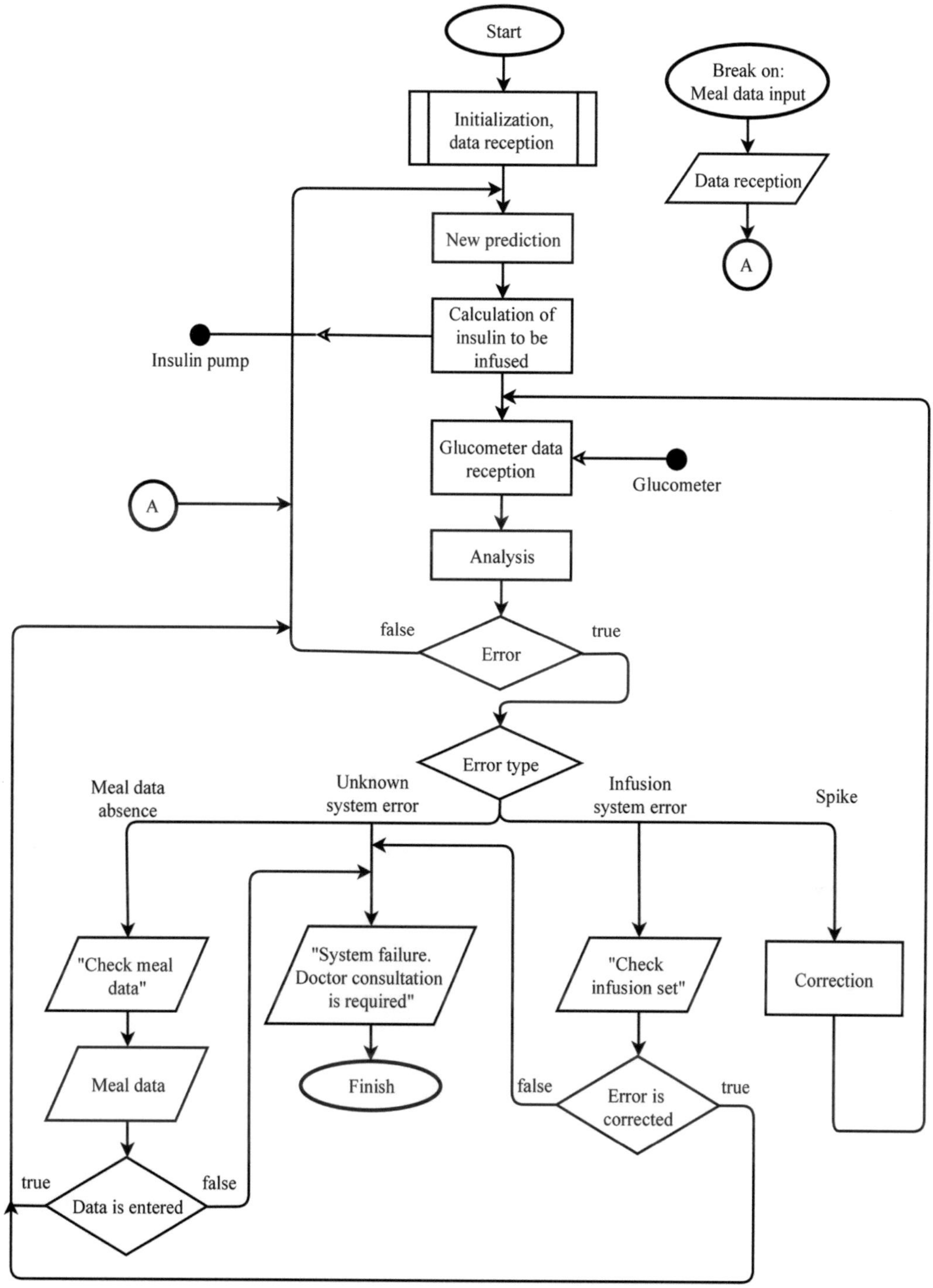

Figure 10. Prediction algorithm operation flowchart.

algorithm confirmed its sensibility to mentioned situations; however, they rarely happen in practice.

Measured	Algorithm	
	Correct	Error
Correct	1102 (96.3%)	42 (3.7%)
Error	4 (2.5%)	152 (97.5%)

Table 2. Algorithm test results.

4. Results and discussion

Sigma model rather accurately describes BG dynamics in patients with diabetes mellitus. For initial model approbation, only food intake and insulin injection data were taken into account. Other factors affecting BG dynamics can be included in the model after additional research. In this case, more precise clinical data will be needed for mathematic model approbation. Nevertheless, the advantage of sigma model is that it does not require a lot of physiological parameters of patient and can be used for BG prediction as is.

Described model approbation shows that RMSE of Bennett-Gourley model is 18.21 mg/dl. In spite of its simplicity, sigma model has the lowest RMSE among all tested models—15.7 mg/dl. This model contains no differential equations and requires minimal computing power in "artificial pancreas" technical embodiment.

Developed predictive algorithm is based on the sigma model. The algorithm estimates the difference between theoretical and experimental tracks with possibility of further correction. BG prediction algorithm is needed for detection of incorrect measurements and correction spikes on BG track. That allows patient to avoid incorrect insulin infusion or glucose intake and reduce possible risks.

Moreover, the algorithm alerts about closed-loop BG control system failures. It allows patient to detect damage or cross clamping of insulin pump infusion set and actualize meal data. Meanwhile patient at his own risk can continue system operation.

The algorithm allows making glucometer-insulin pump system automatic and its work more physiological. Due to insulin doses and injection time calculation, it is infused in advance (before patient BG ascent). Automatic system avoids errors when patient calculates insulin doses and injects it himself.

The algorithm was tested using 52 tracks with full information about the patient's schedule of food intake, insulin injections, etc. Thus, the algorithm suggested in this work makes it possible to avoid 97.5% cases of incorrect operation of the BG monitor. In this case, the RMSE of the calculated blood glucose concentration values was 11.06 mg/dl.

Acknowledgements

This work was supported by the Russian Ministry of Education and Science. Project number 14.578.21.0186. Identifier of the project RFMEFI57816X0186.

Author details

Nikolay A. Bazaev and Kirill V. Pozhar*

*Address all correspondence to: kir-p@rambler.ru

National Research University of Electronic Technology, Zelenograd, Moscow, Russian Federation

References

[1] Cobelli C., Renard E., Kovatchev B. Artificial Pancreas: Past, Present, Future. Diabetes. 2011; **60**(11):2672–2682. DOI: 10.2337/db11-0654

[2] Helton K.L., Ratner B.D., Wisniewski N.A. Biomechanics of the Sensor-Tissue Interface-Effects of Motion, Pressure and Design on Sensor Performance and Foreign Body Response-Part II: Examples and Application. Junior Diabetes Science Technology. 2011; **5** (3):632–646. DOI: 10.1177/193229681100500317

[3] Bergman R.N. Toward Physiological Understanding of Glucose Tolerance: Minimal-Model Approach. Diabetes. 1989; **38**(12):1512–1527. DOI: 10.2337/diab.38.12.1512

[4] Derouich M., Boutayeb A. The Effect of Physical Exercise on the Dynamics of Glucose and Insulin. Journal of Biomechanics. 2002; **35** (7):911–917. DOI: 10.1016/S0021-9290(02) 00055-6

[5] DeGaetano A., Arino O. Mathematical Modelling of the Intravenous Glucose Tolerance Test. Journal of Mathematical Biology. 2000; **40** (2):136–168. DOI: 10.1007/s002850050007

[6] Sturis J., Polonsky K.S., Mosekilde E., Van Cauter E. Computer Model for Mechanisms Underlying Ultradian Oscillations of Insulin and Glucose. American Journal of Physiology. 1991; **260** (5):E801–E809.

[7] Engelborghs K., Lemaire V., Belair J., Roose D. Numerical Bifurcation Analysis of Delay Differential Equations Arising from Physiological Modelling. Journal of Mathematical Biology. 2001; **42**(4):361–385. DOI: 10.1007/s002850000072

[8] Bennett D.L., Gourley S.A. Asymptotic Properties of a Delay Differential Equation Model for the Interaction of Glucose with the Plasma and Interstitial Insulin. Applied Mathematics and Computation. 2004; **151**(1):189–207. DOI: 10.1016/S0096-3003(03)00332-1

[9] Bazaev N.A., Pozhar K.V., Rudenko P.A. Mathematical Modeling of Blood Glucose Concentration Dynamics. Biomedical Engineering. 2015; **48**(6):292–296. DOI: 10.1007/s10527-015-9472-3

[10] Ruedy K.J., Beck R.W., Xing D., Kollman C. Diabetes Research in Children Network: Availability of Protocol Data Sets. Journal of Diabetes Science and Technology. 2007; **1** (5):38–745. DOI: 10.1177/193229680700100519

[11] Diabetes Research in Children Network (DirecNet) Study Group. The Accuracy of the CGMS™ in Children with Type 1 Diabetes: Results of the Diabetes Research in Children Network (DirecNet) Accuracy Study. Diabetes Technology & Therapeutics. 2003; **5** (5):781–789. DOI: 10.1089/152091503322526987

[12] Diabetes Research in Children Network (DirecNet) Study Group. The Accuracy of the GlucoWatch® G2™ Biographer in Children with Type 1 Diabetes: Results of the Diabetes Research in Children Network (DirecNet) Accuracy Study. Diabetes Technology & Therapeutics. 2003; **5**(5):791–800. DOI: 10.1089/152091503322526996

[13] Diabetes Research in Children Network (DirecNet) Study Group. A Multicenter Study of the Accuracy of the One Touch Ultra Home Glucose Meter in Children with Type 1 Diabetes. Diabetes Technology & Therapeutics. 2003; **5**(6):933–941. DOI: 10.1089/152091503322640971

[14] Diabetes Research in Children Network (DirecNet) Study Group. The Effects of Aerobic Exercise on Glucose and Counterregulatory Hormone Concentrations in Children with Type 1 Diabetes. Diabetes Care. 2006;**29**(1):20–25. DOI: 10.2337/diacare.29.01.06.dc05-1192

[15] Diabetes Research in Children Network (DirecNet) Study Group. Continuous Glucose Monitoring in Children with Type 1 Diabetes. The Journal of Pediatrics. 2007; **151**(4):388–393. DOI: 10.1016/j.jpeds.2007.03.047

[16] Diabetes Research in Children Network (DirecNet) Study Group. FreeStyle Navigator Continuous Glucose Monitoring System Use in Children with Type 1 Diabetes Using Glargine-Based Multiple Daily Dose Regimens: Results of a Pilot Tria. Diabetes Care. 2008; **31**(3):525–527.

[17] Diabetes Research in Children Network (DirecNet) Study Group. Relationship of A1C to Glucose Concentrations in Children with Type 1 Diabetes: Assessments by High-Frequency Glucose Determinations by Sensors. Diabetes Care. 2008; **31**(3):381–385. DOI: 10.2337/dc07-1835

[18] Diabetes Research in Children Network (DirecNet) Study Group. Accuracy of Newer-Generation Home Blood Glucose Meters in a Diabetes Research in Children Network (DirecNet) Inpatient Exercise Study. Diabetes Technology & Therapeutics. 2005; **7**(5):675–680. DOI: 10.1089/dia.2005.7.675

[19] Diabetes Research in Children Network (DirecNet) Study Group. Impact of Exercise on Overnight Glycemic Control in Children with Type 1 Diabetes Mellitus. The Journal of Pediatrics. 2005; **147**(4):528–534. DOI: 10.1016/j.jpeds.2005.04.065

[20] Diabetes Research in Children Network (DirecNet) Study Group. Prevention of Hypoglycemia During Exercise in Children with Type 1 Diabetes by Suspending Basal Insulin. Diabetes Care. 2006; **29**(10):2200–2204. DOI: 10.2337/dc06-0495

[21] Tolic I.M., Mosekilde E., Sturis J. Modeling the Insulin–Glucose Feedback System: The Significance of Pulsatile Insulin Secretion. Journal of Theoretical Biology. 2000; **207** (3):361–375. DOI: 10.1006/jtbi.2000.2180

[22] Bazaev N.A., Pozhar K.V., Pletenev A.N. Classification of Factors Affecting Blood Glucose Concentration Dynamics. Biomedical Engineering. 2013; **47**(2):100–103.

[23] Bazaev N.A., Pozhar K.V., Pletenev A.N. An Algorithm for Short-Term Prediction of Blood Glucose Concentration. Biomedical Engineering. 2014; **47**(6):312–314.

[24] Pickup J.C., Renard E. Long-Acting Insulin Analogs versus Insulin Pump Therapy for the Treatment of Type 1 and Type 2 Diabetes. Diabetes Care. 2008; **31**(2):S140–S145. DOI: 10.2337/dc08-s235